The Politics of Swidden Farming

The Politics of Swidden Farming: Environment and Development in Eastern India

Debojyoti Das

ANTHEM PRESS

Anthem Press
An imprint of Wimbledon Publishing Company
www.anthempress.com

This edition first published in UK and USA 2018
by ANTHEM PRESS
75–76 Blackfriars Road, London SE1 8HA, UK
or PO Box 9779, London SW19 7ZG, UK
and
244 Madison Ave #116, New York, NY 10016, USA

British Library Cataloguing-in-Publication Data
A catalogue record for this book is available from the British Library.

Library of Congress Cataloging-in-Publication Data
Names: Das, Debojyoti, author.
Title: The politics of swidden farming : environment and development in eastern India / by Debojyoti Das.
Description: London, UK; New York, NY: Anthem Press, an imprint of Wimbledon Publishing Company, 2018. | Includes bibliographical references and index.
Identifiers: LCCN 2018022207| ISBN 9781783087754 (hardback) | ISBN 1783087757 (hardback)
Subjects: LCSH: Shifting cultivation–India–Nagaland. | Naga (South Asian people)–Agriculture.
Classification: LCC S602.87.D37 2018 | DDC 631.5/8180954165–dc23
LC record available at https://lccn.loc.gov/2018022207

ISBN-13: 978-1-78308-775-4 (Hbk)
ISBN-10: 1-78308-775-7 (Hbk)

This title is also available as an e-book.

This book is dedicated to Georges Condominas (b. 1921) who passed away on 17 July 2011 and who had such a remarkable influence on swidden studies in Southeast Asia (Vietnam); and to the friends of the study village who helped me in every possible way during my fieldwork; and to our loving son, Aryan.

CONTENTS

ILLUSTRATIONS

Figure

Images

Maps

Tables

FOREWORD

The title of this book, *The Politics of Swidden Farming: Environment and Development in Eastern India*, hardly conveys the ethnographic depth and historical reach of the study. This scope becomes apparent as the reader comes to discover the significance of swidden or shifting cultivation, not just as a relationship of communities to land and ecology, but as a critical mediator of relationships with the colonial and postcolonial state and its civilizing or development projects over the past century. The study is located in a region and among people who become, through their social lives and livelihoods (including their swidden cultivation), as Anna Tsing puts it, 'icons of the archaic disorder that represents the limit and test of state order and development' (Tsing 1993: 28). The remote Naga villages that are the ethnographic focus of this book are in this sense represented simultaneously as a political–administrative border, an agricultural margin and a moral frontier. The kinds of colonial and postcolonial projects that have sought to regulate and control such borderland communities, whether irrigation projects, roads, settlement schemes or plantations, are well known, but what is rarer is the kind of longitudinal account offered here of the interplay of state power, Christian mission and local communities, examined also through the complex relationships between and within these Nagaland villages themselves. With regard to this latter theme of the role and agency of local communities in their own transformation, it becomes evident not only that colonial administrative power was asserted through harnessing inter-group antagonisms, but also that transformations brought to the cultivated landscape, to social institutions and cultural practices through Christian missions, came by way of neighbouring social groups, who modelled new ways of worshipping and working the land, expanding cash-crop or rice cultivation (an established archetype of settled civility).

Violence is a thread that runs through the account of the political and moral relationships of this frontier zone. Debojyoti Das offers the reader his own experience of being caught up in a violent episode involving a kidnap, interrupting the decade-long ceasefire between the army and Naga militants, as a route into ethical and methodological reflection. As the book explains, through use of remarkably candid archival sources on the brutality of the British administration, the people of this frontier have long existed as the target of counter-insurgency violence, both physical and representational. Having imagined the local Naga communities in terms of their violence – as head-hunting, slave-taking savages and insurgents – British 'punitive expeditions' of quite remarkable destruction and violence were undertaken. The moral framing of such violence as civilizing allowed an unguarded record of the dreadful violence involved in such penalizing action. All the more disturbing for the scholarly discipline in which Das is trained, is the evidence he brings to show that it was in alliance with this administration that anthropology's 'rescue and record' ethnography was produced, including the photographic capture of the 'naked Nagas' for display in the *Illustrated London News* from the 1930s. Intersecting violence meant the British frequently exploited existing inter-group rivalries, and in the context of World War II, actually rewarded Nagas for the taking of Japanese heads, the practice that had justified their punitive expeditions.

Das has discovered a wealth of new archival material on the colonial administration of this frontier region kept in local record offices. Despite the important light thrown on direct military assertions of colonial power, these sources are used to show how in practice the colonial administration was an outcome of negotiation and depended significantly upon the recognition of local rights in land, and that it was through 'customary practices' that the government control over land was legitimized. This turned village heads into brokers or middlemen in the extension of commercial forestry, and the transformation of swidden (*jhum*) land with new rotations and alder pollarding, alongside the expansion of permanent wet-terrace rice farming. A new regime of stabilized land tenure went along with and allowed new farming practices. We learn that the familiar story of swidden as a farming system pressurized to the point of collapse with increasing population and shortening fallow cycles is a simplification, and with case-study detail we come to understand the remarkable adaptability of such land-use practices to changed circumstances.

Despite the organizational force of colonial administration, one of the surprising lessons is that perhaps the greatest change in the village community Das lived in was wrought through self-transformation initiated by Christian (primarily Baptist) missionization by evangelists from neighbouring rice-growing communities. Many accounts of the evangelization of Indian communities show Christian adaptation to existing cultural and agricultural systems, as church festival and ritual systems are moulded to an existing popular religious culture to produce forms of religious synthesis. In this account of

Nagaland, by contrast, conversion to Christianity brings new order and discipline that involve the transformation of cultural and agricultural systems – a disinvesting from the old sources of social power and investing church-based institutions with authority, including in matters such as labour control as well as connecting outwards to aspects of Naga nationalism's cultural forms. Time and its reorganization in terms of calendrical systems turn out to be a critical linking medium of religious, social and agricultural change. New agricultural practices, especially irrigated rice, become stores of symbolic capital and social differentiation.

It is perhaps unsurprising that development-project planners with characteristic lack of historical sensibility and the tendency to dichotomize the recent traditional past and the technologically improved future fail to perceive the dynamic systems into which their interventions are introduced. And so it is with the Nagaland Empowerment of People through Economic Development (NEPED) project, which for Das first encapsulated the conundrum of swidden farming. Project goals of forest regeneration to produce 'carbon sinks' and to counter global warming are realized through incentivizing swidden farmers – simultaneously the most knowledgeable custodians of local biodiversity and its principal threat – to grow horticultural and plantation cash crops and thereby achieve *sustainable* swidden farming. But contrary to expectations, the transition from swidden to agroforestry does not bring livelihood improvement for all. Instead the development project is another chapter in the long story of intersections between diverse local interests and outsider ambitions for cultural–agricultural change; and in this case new opportunities are found to encourage forms of development brokerage, elite capture and the privatization of benefits, which Das uncovers.

Few accounts of remote borderland regions show with such attention to historical and local contexts how technologies of cultivation and technologies of rule are interconnected; how marginality is reproduced through regimes of development, alongside internal differentiation and tension within communities. The story reveals how the interweaving of changes in political, economic and religious relations can only properly be understood by paying attention to what is happening in the landscape; and simultaneously how the landscape must be read as a palimpsest of political and cultural histories.

It takes sensitive development of relationships with communities through long-term fieldwork, an arduous searching out of local records and informed use of a visual archive of colonial ethnography to arrive at such understanding. This in turn has to be treasured for the lessons it offers to current and future projects in such frontier regions, allowing more informed critical awareness of how developers' enduring notions of civility and proper citizenship come to be expressed in the idiom of improved forms of land use and farming. And at the very centre, and representing the ambiguity of such development, is swidden cultivation balancing between visions of primitive destruction and sustainable ecology that are part of contemporary environmental discourses.

David Mosse, FBA
London, April 2018

ACKNOWLEDGEMENTS

Scores of people and many institutions have been helpful in the preparation of this book, and I wish to acknowledge their assistance. I shall remain ever indebted to their sincere help and generous goodwill. It is through their constant encouragement and support that I was able to complete my enduring field research amid medical problems, anxiety with settling down in the field villages and the challenges I faced while doing fieldwork. I would like to thank my advisor David Mosse for his expert guidance, stimulating discussions and practical instructions for my research and writing up despite his family bereavement and research leave. I would also like to thank my second supervisor Johan Pottier (an Africanist) for his encouragement and help at every stage of the work.

In the study village I was well supported by my host family and the village people. I would like to thank my host's daughters and his nephew, who helped me in my interviews and made me understand the social geography and clan dynamics of the village. The pastors and women leaders were of incredible help and I enjoyed passing the time talking about their daily life and often cracking jokes. I would also like to convey my regards to the village 'citizen president', the second *gaon burha*, and other headmen who took me to their *jhum* and paddy fields and helped me to understand the complicated issues surrounding farming, land use and social relations. I am also thankful to the Catholic father of Shamator town who hosted me on several occasions when I could not go back to the village after dark. Also, the head *dobashi* and head *gaon burha* of Shamator town shared invaluable information on administration and colonial times. Similarly, I am indebted to the help extended by the Kecham family, who hosted me in their house.

I am indebted to my parents and younger brother for having taken pains to periodically send articles, books and periodicals published from the region. I also thank my relatives, friends and members of my extended family who hosted me during fieldwork trips between Assam and Nagaland.

As for the various institutions that gave research support, I am grateful to the Nehru Memorial Museum and Library, the National Archives and the

Central Secretariat Library in New Delhi; the Nagaland University, ICSSR-NER Shillong Library, North Eastern Hill University and North East Council Library and the Keeper of Records of the Meghalaya State Secretariat in Shillong; the Asiatic Society of Bengal; the Assam State Archives in Dispur, Guwahati; the West Bengal State Archives in Kolkata; the Arunachal State Archives in Itanagar; the Department of Antiquarian Studies; the Department of Art and Culture–Record Cell, Nagaland, Kohima; the Nagaland State Secretariat and the Naga Student Federation (office records) in Kohima; the Clarke Theological College (CTC) and Deputy Commissioners Office Records in Mokokchung; the Village Guard and Land Survey Department, Agricultural and Flood Control Department and Deputy Commissioner's office in Tuensang; the Nagaland Baptist Church Council (NBCC) Library in Guwahati; the School of Oriental and African Studies Library (SOAS) and Christoph von Fürer-Haimendorf Special Collection, JICS Digital Archives, the Royal Anthropological Institute Archives, Anti-Slavery International, Kew Botanical Garden Archives and the British Library in London.

Finally, I gratefully acknowledge the various institutions that helped me with financial support. I am particularly thankful to the Felix Foundation that supported my PhD research for three years (2007–10). Similarly, I would like to thank the trustees of the Christoph von Fürer-Haimendorf Anthropological Fund for awarding me a fieldwork grant. My special thanks to Nick Haimendorf, who took time out of his busy professional life to attend my departmental post-graduate PhD seminar presentations. I thank the University of London and SOAS for their fieldwork grant, and the Royal Anthropological Institute (RAI) for awarding me the Emslie Horniman Anthropological Research Grant to buy equipment (a camera and GPS with which I took beautiful pictures in the field, measured altitude and determined coordinates).

My research was also supported by bursaries and writing-up grants. I would like to thank the trustees of the Leache Trust, the Gillian Diamond Trust, the Nora Henry Trust, the International Student House Hardship Fund, Churches Together of England, the Economic and Social Research Council (ESRC), the Radcliffe-Brown Sutasoma Award, the Royal Anthropological Institute and the Sydney Parry Foundation for their crucial support.

I am most indebted to my Yimchunger Naga friends and companions whose names appear in pseudonym for obvious reasons (matters of confidentiality). This thesis has been a result of teamwork not only between my interlocutors and me, but also between the various institutions, individuals, *jhum* farmers, civil society members and church office bearers and academics who participated and made this research possible. Also, thanks go out to a few academic mentors and friends who have been a source of inspiration throughout the writing process: Vinita Damodaran (Sussex University),

Edward Simpson (SOAS), Ambika Aiyadurai (NUS), Sunil Amrith (Harvard University), Amit Kisku (SOAS), Niranjan Saha (SOAS), Raminder Kaur (Sussex University), Andrew Ainsley (Reading University), Barbara Harris White (Oxford University), Mandy Sadan (SOAS), Willa Zhen (CIA), the late Bianca Son (SOAS), Kuntala Lahiri-Dutt (ANU), Keerty Nakrey (Queens University), M. Satish Kumar (Queens University), Rohan D'souza (Kyoto University), Antonio Moon (British Library), Farzana Whitefield (SOAS), Sahil Warsi (SOAS), Vibha Arora (IIT, Delhi), K. Sivaramakrishna (Yale), Mahesh Rangarajan (Ashoka), Erik De Maaker (Leiden University), Angela Chiu (SOAS) and George Kunnath (Oxford University).

For their editorial support I would like to thank Kieran Nelson, Baverley Brown and the copyediting team at Anthem Press, who carefully edited the manuscript to flawlessness. My special thanks go to the commissioning editors. The shortcomings in this book are entirely mine. And for copyright clearance I would like to specially thank SOAS Library, the *Illustrated London News*, Pitt Rivers Museum – University of Oxford, and Nick Haimendorf for allowing me to reproduce the photographs and works of Christoph von Fürer-Haimendorf and J. P. Mills.

Lastly, I shall not forget the service rendered by the University College Hospital (UCH) team of doctors in the Department of Respiratory Disease who helped me recuperate from my medical condition, tuberculosis, which temporarily endangered the prospects of my preparation for field research. I am also thankful to Pie Chen, Max Mazamdar and Ian Harper at Edinburgh University for their valued counselling and support.

ABBREVIATIONS

ADG	Army Development Group
AFSPA	Armed Forces Special Powers Act 1958
AIPP	Asia Indigenous Peoples Pact
ANSAM	All Naga Students' Association Manipur
BRO	Border Roads Organisation
CCP	Chinese Communist Party
CNBC	Council of Naga Baptist Churches
DAN	Democratic Alliance of Nagaland
DB	*dobashi*
DC	Deputy Commissioner
ENPO	Eastern Naga Peoples Organisation
FGN	Federal Government of Nagaland
GB	*gaon burha* – village headman
GOI	Government of India
GPRN	Government of the People's Republic of Nagaland
GPS	global positioning system
IAY	Indira Awaas Yojana (Indira Awaas Yojana has been renamed as Pradhan Mantri Gramin Awaas Yojana by the NDA government)
IAITTF	International Alliance for Indigenous and Tribal Peoples of the Tropical Forests
ICEF	India Canada Environmental Facility
ICITP	Indian Council of Indigenous and Tribal Peoples
IDRC	International Development Research Centre
IIL	Indian Independence League
ILP	inner line permit
INA	Indian National Army
KWIA	Flemish Support Group for Indigenous People
MCA	Military Civic Action
MGNREGS	Mahatma Gandhi National Rural Employment Guarantee Scheme

MRG	minority rights group
Naga Hoho	Apex organization representing Naga communities
NBCC	Nagaland Baptist Church Council
NEPED	Nagaland Empowerment of People through Economic Development
NHDTC	Naga Hills District Tribal Council
NHTA	Naga Hills Tuensang Area
NMA	Naga Mothers Association
NNC	Naga National Council (Adino's group), associated with the signing of the Shillong Accord
NNO	Naga National Organisation
NPCC	Nagaland Pradesh Congress Committee
NPF	Naga Peoples Front
NPMHR	Naga Peoples' Movement for Human Rights
NSCN	National Socialist Council of Nagaland
NSCN–IM	National Socialist Council of Nagaland – Isak, Muivah and Khodao
NSCN–K	National Socialist Council of Nagaland – Khaplang
NSF	Naga Students' Federation
NWUM	Naga Women's Union, Manipur
PAP	protected area permit
POU	project operation unit
RAP	restricted area permit
RM	resource mapping
RMO	Range Management Office
SACON	Salim Ali Centre for Ornithology
SEAGST	Southeast Asia Graduate School of Theology, Singapore
SHG	self help group
UDHR	Universal Declaration of Human Rights
UN	United Nations
UNC	United Naga Council
UNDRIP	United Nations Declaration on the Rights of Indigenous Peoples
UNESCO	United Nations Education, Scientific, and Cultural Organisation
UNPFII	United Nations Permanent Forum on Indigenous Issues
UNPO	Unrepresented Nations and Peoples Organisation
VC	village council
VDB	village development board
VG	village guard
WGIP	Working Group on Indigenous Populations

1

Introduction

> Both colonial technical staff and farmers in the plains show little consideration for and even despise swidden practices all over their colonies in Southeast Asia. Yet one hears little of the centuries during which the Mayan civilization ruled over Central America, which is strewn with their masterpieces. The colonial administration did not know, for example, that swidden had played a major role in the demographic expansion of the twelfth century in France: it transformed vast areas into cultivated fields of cereals. It was also the case that swidden was widespread in Europe during the nineteenth century, and it is also mentioned as existing in Austria during the 1960s. If you walk through the forest of Fontainebleau you will come across many sites or hamlets with the name 'I'Essart' or 'Essart' (swidden) on the survey maps. (Condominas 2009, 267)[1]

The seed of this book dates back, at least in part, to 2006, when I attended a seminar on 'shifting cultivation' – swidden farming, pejoratively known as 'slash-and-burn', and in northeast India as *jhum*. This seminar, delivered by P. S. Ramakrishnan, inspired me to undertake my research among the hill farmers of Eastern India. This manuscript is a labour of love, written through years of commitment working with *jhum* farmers in Nagaland. Ramakrishnan was, at that time, a member of the advisory committee in a transnationally funded *jhum* regeneration project. The project, entitled Nagaland Empowerment of People through Economic Development (NEPED), was funded by the Canadian International Development Research Centre (IDRC) and the India–Canada Environmental Facility (ICEF) primarily to implement 'carbon sink' and target global warming and climate change–induced environmental risks through the regeneration of forest land and by incentivizing farmers to grow horticultural and plantation cash crops in the eastern Himalayas. This region is a major biodiversity hot spot in South Asia and has grasped the attention of biologists, geographers, climate scientists and ecologists who see *jhum* as a hazardous and unproductive form of farming, threatening local biodiversity.

The appraisal of the project's success and research funding from the Felix Scholarship, UK, to pursue my work at the Anthropology Department of the School of Oriental and African Studies drove me to hike the Saramati mountain range across the India–Myanmar borderland. I lived among the Yimchungers to develop my field-based knowledge, which informs the ethnography in the book.

In this book, I trace the story of agroecological change and state intervention back to colonial times when the Naga Hills were seen as the frontier of state and 'civilization'. We need to understand contemporary agrarian change by contextualizing farming, not just in terms of the science and technology of agriculture or conservation and biodiversity, but also in terms of 'technologies of rule'. For the colonial administrators of the Naga Hills – who saw their role partially in terms of rescue-and-record ethnography – *jhum* practices were part of backward Naga customs and traditions. 'Improving' farming practices was bound up with indirect rule as a distinct process of governance involving forms of knowledge and intervention. It was political expediency rather than 'imperial science' that changed local agroecologies and put pressure on the practice of shifting cultivation. Crucially, neighbouring Naga terrace rice cultivators were promoted as offering a more civilized – yet local – alternative.

Here I propose to deliver anthropological insights into the social ecology of farming that may help to understand how swidden has been framed within statist and epistemic discourses that have informed both policy and practice on agriculture development in upland South Asia. Besides developing an ethnography of *jhumland* development in the hills, the work has other objectives that originate from my engaged and collaborative ethnographic fieldwork and from my use of multiple sources and methods of data collection in a sensitive field site (particularly colonial photographs, archival records and policy dossiers). This work also demonstrates how contemporary agrarian development reflects this complex colonial heritage, including linkages between the state and village elites. Evangelical missionaries in the post-independence period also contributed by appropriating local institutions and incorporating them into a Protestant (Baptist) ethic of work. Reinforcing the colonial state's favouring of rice as the 'crop of civilization', the missionaries' moral discourse installed new time disciplines geared to settled agriculture.

Methodologically, I engage with the many voices that shaped my field research, providing evidence from in-depth, household-based participant observation and life histories and a household survey, while also drawing extensively on original archival research and colonial photography to provide documentation of colonial representations of the swidden landscape. This research was undertaken in a milieu of fear and violence, which raises further methodological and ethical issues in the book that are relevant for ethnographies carried out in dangerous field sites (see Chapter 2).

Global Discourses: An Overview of Swidden

To begin with I provide an overview of *jhum*, northeast India's swidden farming. Swidden agriculture is a technique of rotational farming in which land is cleared for cultivation (normally by fire) and then left to regenerate after a few years. This is followed in a cycle. However, in northeast India, as elsewhere in the tropics, various factors – such as demographic changes, the introduction of new cash and plantation crops, the building of big dams and the submergence of forest land in catchment areas – have reduced the land use needed for swidden farming. Simultaneously, governments worldwide have long sought to eradicate swidden agriculture, terming it 'slash and burn' because of an erroneous belief that it is the sole driver of deforestation and soil erosion in the hills.

Swidden is today increasingly understood in national and transnational agroecological discourse as an obsolete form of land use that not only puts pressure on land and its dwellers, but also destabilizes forests, soil and biodiversity in highland ecosystems. The environmental narratives of shifting cultivation that induced damages to ecology were produced, and scientifically deliberated, during the early nineteenth century and later reaffirmed in colonial forest policy during the British empire's scientific conferences held between 1921 and 1951 (Rajan 2006). Since the 1950s, soil erosion and desiccation narratives have joined neo-Malthusian discourse on the 'carrying capacity of land' to discourage swidden farming. More recently, revisionist scholars have critiqued scientific claims about the negative impact of swidden by looking at national programmes of territorialization that historically marginalized the forest-dependent swiddeners and separated them from their livelihood. Among the revisionist school, neo-indigenistas argue that swidden farming is sustainable when practised under long cycles of field rotation (Agrawal 1995). They cite the cultural attributes of swidden that make it impossible to be abandoned completely, leading to the failure of many *jhum* improvement schemes of national governments and transnational conservation organizations that deny this reality. Agriculture, as the neo-indigenistas claim, is inherently a cultural practice, and if communities are alienated from it, the process of institutional intervention through new technologies and incentives for improvement will fail. On the other side of the political scale, the radical camp of 'political ecologists' base their critique on the economic and political circumscription of swidden populations by national policies that favour big capital and transnational timber cartels (Dove 1983). Taking into account the points of view and contemporary critiques on *jhum*, I argue for a 'cultural ecology of conservation effort'. I will explain through a village-level case study the less-understood histories of state territorialization of swidden landscapes, ideas of modernization brought in by missionaries, changing land use and labour relations and shifting values towards access and rights to land under 'community ownership', along with the rhetoric of population implosion in swidden villages. Jointly, these developments have transformed swidden landscapes from their ideal beginnings as self-sustaining 'little insular-barter economies' into timber and cash-crop plantations that sustain export to urban and regional markets on the plains. In a rapidly changing globalized rural situation, swidden farming is hybridized with the introduction of agro-forestry, silviculture and timber-plantation economies overlaid by traditional land-use systems. Villages are vertically integrated to the market and to trans-national capital. Equally, swidden farming performed under state surveillance is territorialized, fragmented, settled and regulated by forest and wildlife protection laws as well as by new schemes of permanent husbandry. The modern swidden economy has to be understood as an agroecological and cultural system under transition and shaped by epistemic discourses, state and trans-national development promises and programmes of conservation as well as the swidden communities' 'will to improve' (Li 2007) – a desire to catch up with the outside world.

This book is based on my ethnographic case study of the Yimchunger Naga tribal group, a minority within the pan-Naga and Indian identity. They remain one of the least documented of the Naga tribes bordering India and Myanmar. Their marginality is evident both in their geographical and political relationships with the lowland (plains caste) population and with the Indian state: theirs is a region that rarely figures in Indian Studies (Kejariwal 1987, Maaker and Joshi 2007, Kumar and Zou 2011). The region's farming and landscape can best be compared with the mountain in Southeast Asia that James C. Scott has generalized as 'Zomia' (2009: ix).[2] The central purpose here is to test the simplifying discourses of swidden agriculture as 'underdeveloped, backward, waste and marginal' which, after being initially established in the early nineteenth and twentieth centuries within colonial rescue-and-record evolutionist Naga ethnography (Lotha 2007), later became incorporated in the process of state planning. By contrast, I hope to demonstrate the dynamic nature of swidden agriculture among the Yimchunger Nagas, while also tracing the historically specific social and political factors that shape this land-use practice.

I narrate here, in brief, my entry to the field, which will serve as a vignette to the perceptions on swidden, or shifting cultivation. In February 2009, I made a trip to Tuensang town via Mokokchung with a friend who worked

Map 1.1 Location map of Nagaland and the study region.

as the principal in a government college. The journey to Tuensang was memorable, and in the coming months it became more regular once I established my fieldwork base among the Yimchunger Nagas (a Naga subtribe inhabiting the Tuensang and Kiphire districts of Nagaland; see Map 1.1). The tribe's members offered insights into their agricultural methods, my subject of enquiry. During the journey, I encountered the natural beauty of the landscape, spotted with dark, charred hills that appeared regularly as we ascended the Saramati peak. This landscape serves as a natural watershed between India and Burma. Here there were newly opened *jhum* plots prepared for the coming season. A tribe member was quick to present his views of the *jhummed*

Image 1.1 A climate-change poster at Clerk Theological College, Mokokchung.
Source: Author's photograph, 2009.

landscape amid dense subtropic greenery. 'Look at these *jhum* plots, people here are mostly dependent on slash and burn. They make much bigger *jhum* fallow than in other parts of Nagaland. They are still backward in their agriculture. The landscape will soon be robbed of fertile soil by the monsoon showers.' My interlocutor soon followed with the remark, 'We Nagas still appreciate it as a way of life.' The practice of *jhum*, despite being embedded in people's culture, brings immeasurable stress to life and landscape in the region. These remarks come from every government official, non-governmental organization (NGO) worker, missionary, schoolgoing teenager and often from farmers themselves (see Image 1.1). *Jhum* is presented as a bane, and the farmers express their helplessness in light of limited opportunities to transform their agrarian livelihood, which is dependent on slash and burn. In this book I take these perceptions of *jhum*, based on 'received wisdom' (Fairhead and Leach 1996), and situate them within a wider historical and ethnographic context of colonial and missionary contact that framed the *jhum*-based Naga livelihood as part of the frontier wildness and as being ignorant and savage-like. This anthropological–historical lens, focused on the changing life and times in Naga society, will help understand how we imagine agrarian change and farming in Nagaland, on the one hand, and state–Naga relations on the other.

The points are crucial to unveil the nature of our contemporary knowledge about swidden cultivators and their lived world. During my field research in Nagaland (2008–9), I recorded several accounts of how outsiders (town people, plains people, project staff, government officials) who no longer practice *jhum* described it as bad, wasteful, primitive and at the same time intrinsic to local culture, traditions and local knowledge. The dual perception of *jhum* as a cultural practice, a way of life and a marker of backwardness led me to ask what is involved in shifting cultivation and why it is backward, wasteful, primitive – all negative connotations attached to a way of life that is quickly changing, along with changing food choices, culture, customs, accessibility to markets and increasing mobility to towns and cities.

Since the decolonization phase of the 1950s, there has been active engagement by scholars in defining *jhum* cultivation, as it became a systematic focus of institutional intervention carried out by the Food and Agricultural Organisation (FAO). The FAO staff, in their attempt to replace swidden with more civilized forms of settled agriculture, presented a narrow and straightforward definition: 'Shifting Cultivation is the custom of cultivating, clearings scattered in the reservoir of natural forest (forest or grass woodland) and of abandoning them as soon as the soil is exhausted' (FAO staff 1957). A negative stereotype of swidden agriculture was developed that would shape policies by national governments in order to 'emancipate' swiddeners of their 'backwardness'. Following this FAO definition, geographers, anthropologists, sociologists and cultural-studies academics working on highland agronomy and indigenous cultures have tried to define this complex farming practice by combining various aspects of *jhum* rotation, burning and fallow management.

However, the definition of swidden is still partial, and no universal categorization can be made of the diverse practices involved. Both Conklin (1961) and Spencer (1966) have proposed a very detailed classification of shifting cultivation from their study among swidden cultivators in Southeast Asia. They have made distinctions, such as coining the term 'linear shift' as opposed to 'cyclic shift', and have pioneered and established categories. In practice, however, these distinctions partly explain the dynamic nature of swidden systems. In fact, Mertz et al. observe that 'swiddeners in Southeast Asia have been engaged in a long-term transition from pioneer to established systems as their livelihoods are transformed' (2009: 261). The increasing speed of change in recent decades has given rise to a diverse array of land use and livelihood systems that defy simple definition.

For example, Rambo (1998), in his study of the highland communities of Vietnam, defines 'composite swiddening' as an agriculture system that integrates permanent wet-rice fields and rotating swidden plots into a single household system. A similar scenario of composite swiddening can be found practised by Angami and Chekhesang Nagas in the Khomia and Phek districts of Nagaland, and this is deemed to be based on well-established *jhum* fallow management (alder tree pollarding and *jhum* rotation) followed along the hill slopes, while terrace farming dominates the inter-mountain and rivulet valley slopes. Cairns and Brookfield (2011: 57–79), based on their analysis of field-work data, identified alder-based pollarding and wet-terrace rice cultivation practised by Angami Nagas in Khonoma and Tsiesama villages in Nagaland as another form of composite farming (but which was changing due to the introduction of cash crops by the Nagaland Empowerment of People through Economic Development (NEPED) programme in the late 1990s and the new opportunities the villagers discovered through government white-collar jobs in towns and crop-diversification opportunities. Likewise, in many parts of Nagaland, intensive agriculture has led to extended cropping over several years in the same location through tree plantation, horticulture and crop combination that extend the fallow period. As I discuss in the book, the Yimchunger Nagas in the field-study village (Leangkonger) have evolved a long bean (*kholar*) relay swidden cycle that has led to an extended *jhum* cycle, which in the eyes of outsiders has often been seen as sustainable. Similarly, Mokokchung district, according to state Agriculture Department statistics, has the most sustainable *jhum* fallow cycle, which is 20 years long. Most of the households no longer practise slash and burn, but have developed plantations, horticulture, small-scale tea gardens and vegetable farms that have ensured permanent tenure. In this context, the definition of swidden cultivation as proposed by the journal *Human Ecology* (2009) is helpful in understanding the changes that have come about in the swidden cultivation landscape in Nagaland, as elsewhere in the upland tropics. 'Shifting Cultivation' is defined as

> a land use system that employs a natural or improved fallow [in] the cultivation phase of annual crops, sufficiently long to be dominated by woody vegetation, and cleared by means of fire. The staple food is most often upland rice, but can be maize […] secondary crops such as cassava, banana and other annual or perennial crops occur to varying degrees in the swidden as do cash crops such as ginger, cardamom, etc. (Mertz et al. 2009, 261)

Shifting cultivation, or slash and burn, is semi-nomadic crop husbandry practised by people who are semi-nomadic and produce very little surplus for market transactions, a problematic agrarian method practised by ethnic communities in the highlands. In the Northeast of India as well, scholars have defined shifting cultivation following the global lexicon as a challenging land-use system, yet also view it as an indispensable way of life. Only a few scholars like P. S. Ramakrishnan (1992) have given a positive framework to understand *jhum* as a sustainable system. However, even he subscribes to the neo-Malthusian view that ascribes *jhum* cultivation vulnerability to population pressure, specifically the idea that this is the sole cause of its non-sustainability in present times. The neo-Malthusian approach has been highly influential in defining *jhum* as non-sustainable in terms of 'purely' technical reasoning – 'purely' meaning ignoring the pressures on *jhum* imposed by the political economy of land appropriation, enclosures and reserves under state projects of modernization that alienate swidden farmers from their livelihoods (Dove 1993). Looking at the problem as it emerges from colonial and Western discourses, which narrow swidden down to 'slash and burn' and define it as a negative practice, I shall begin by explicating how it is named.

What's in a Name?

The common term slash and burn is implicitly pejorative and generates a powerful stereotype. Slash and burn offers a potent combination of images of fire, forest destruction and blackened landscapes. When the colonial administration first encountered the shifting cultivators in the Naga Hills, their immediate reaction was one of derision for the wildness and 'waste' of *jhum* practices. Plains people who lived in the areas bordering the hills also shared these views. The shifting cultivators, when measured against peasants from the plains and caste Hindu neighbours, were classified as people of 'another time'. Van Schendel (1992) and Guha (1999), in their illuminating work on highland societies in South Asia, have theorized this process as 'ethnic involution-ethnogenesis' or hierarchies through which the colonial government framed the measures of civilization into tribes and peasants. In the Naga Hills colonial administrator-ethnographers were engaged in a form of 'rescue and record' ethnography of local customs. Material culture in the form of physical objects and architecture (e.g. *murung* houses) was highly valued – and farming practices were thus located within this picture. Once collected, objects were kept in museums as archaic emblems of their primitive past. The Pitts Rivers Museum, Oxford, became a major centre for the display of a Naga material culture collection sent by colonial administrator-anthropologists in the Naga Hills, J. H. Hutton and J. P. Mills, all carefully arranged in a timescale (genealogy tree) order by its first curator, Henry Balfour. The primitiveness of the Naga people and their social life was thus made visible and 'real' through these objects of material culture.

Jhum, repackaged as slash and burn in the English-speaking world, is represented as a graphic destructive practice. Whether pictured as fire and destruction or metaphorically fossilized as an emblem of another time, such discourses hide the productive element and negate the dynamic, adaptive and cultural qualities of *jhum*, as these systems work within particular environments. Although seemingly neutral and descriptive, such discursive constructions are highly loaded with values and circulate in relation to wider social and cultural contexts: hills versus plains, productive versus wasteful, civilized versus savage. Such language is not new, but reiterates the colonial tropes of primitive representations of Naga agriculture.

Christoph von Fürer-Haimendorf's observation on *jhum* is particularly revealing:

> Primitive husbandry is certainly crude when gauged by modern standards. It is 'predatory' in the sense that there is little, if any systematic practice of crop rotation or soil compensation. The consequence is naturally exhaustion of soil, involving removal to a new plot, where a new clearing is made with much wastage of timber and energy [...] The development of agriculture techniques by the adoption of the plough paves the way for large-scale agriculture operations and surplus production. It emphasises the permanence of field and settlement and individual claims of specific tracks of land, in other words it encourages private property rights and the growth of permanent villages.[3]

The language used is deeply implicated in an evolutionary and dualistic developmentalist narrative that assumes a progression from nature to culture, from nomadic grazing to sedentary agriculture. On one level there is the inflationary language of destruction, although here fire is replaced with predation. On another level, as in Pelzer's (1945) classification of shifting and settled agriculture, there is a hierarchy of 'advanced' versus 'primitive' farming, in which the former is associated with 'change' and the latter with 'continuity' and stasis. As O'Brien spells out, 'in Pelzer's definition advanced agriculture and culture are presented as "high stage" as opposed to "low stage" forms, characterized by "intensive", "sedentary" versus "extensive," "shifting" agriculture, and practised with "sophisticated" versus "crude" technologies' (2002: 491). The subordinate state in each of these dichotomies implies a less developed and more original (i.e. closer to nature) state of humanity. The modernization transformation that Pelzer desired has remained a dominant goal of agroecological development since the colonial period and of contemporary development, despite, and sometimes because of initiatives to incorporate local knowledge in development.

Jhum cultivation was not only considered closer to nature but a land-use system that had more 'waste uncultivated land'. According to Locke's *Two Treatises on Government* (1690), uncultivated land was empty waste; it was considered as land left wholly to nature, where no property rights could arise because no labour had been 'mixed' with the soil (Whitehead 2010: 92–93). It is not surprising, therefore, that when Europeans encountered tropical lands they did not recognize the culture of shifting cultivation but saw only disorderly 'wastelands'. In Assam, as Baruah (2001) observes, shifting cultivators were all relegated to the margins once colonial land-use policy, framed through the Wasteland Grant Rules of 1838 and the *Zamindari* and *Ryatwari* system, declared the commons, where people *jhummed*, as 'wastelands'. Some of these wastelands were declared as forest reserves, leased by the English tea planters and the Assam Bengal Railway Company. Demand for land intensified through decennial tenure rights given to immigrant wet-rice cultivators who were brought from East Bengal to cultivate wastelands and generate revenue for the Assam administration.

This policy led to the creation of private property in colonial Assam. The colonial capitalist enterprise soon took over the wastelands, and the commons were all settled as productive tea plantations and private landholdings with permanent tenure rights that could not be alienated or kept as *jhum* fallow. The 'mission' of 'civilizing' guided colonial practice of settlement control, domestication and improvement, and efforts proceeded to curtail *jhum*.

Social and political motivations were also important factors behind colonial efforts to prevent shifting cultivation. Endeavours to ban or limit shifting cultivation derived in large part from the desire to impose taxes and secure labour and raw material for the colonial enterprise (plantation, scientific forestry, foot soldiers and coolies). Mobile populations were difficult to tax, and they could not be pinned down to provide labour for colonial activities. Thus, swiddeners in colonizers' minds were irregular, unmapped and non-state subjects; a characterization that Scott (2009) uses to define swidden cultivators as stateless people in his book *An Anarchist History of Highland South East Asia*. As swidden cultivators were prevented from entry into state forests by colonial laws, the forest appeared as a purely natural place. The declaration of 'reserve forest' became an important site of state regulation and power (Puleso and Vandergeest 2001: 761). Both forest and fields were redefined to support and justify colonial resource regimes.

Once tea plantations were established along the Assam-Naga foothills, with indentured labour brought from Chotanagpur and the surrounding regions of Central India as the main workforce, the tea planters came into direct dispute with the Naga villages in the foothills. Simultaneously, the formation of a separate Forest Department in the 1870s in Assam brought the department into competition for wasteland grants with the tea planters: the Forestry Department sought these grants for the purpose of establishing reserves that would then be used for the scientific regeneration of teak and sal (Saikia 2011).

The arrival of the new immigrant labour force in tea plantation and slow advances of tea gardens towards the foothills alarmed Naga swidden farmers and this led to pillaging of plantation coolie colonies by the Nagas. The fundamental change, as Baruah (2001) highlights, was a shift in resource regime from the *jhum* commons, which were now demarcated as reserves, to enclaves of private property with fixed propriety. The result was bloody skirmishes between the Naga swidden cultivators and the tea plantation workers. The Nagas feared their rights to *jhum*, hunting wild animals and the collection of forest products were threatened by the enclosures set up by the tea planters.

'As one mode of resource use comes into contact with another mode organized on very different social and ecological principles', write ecological historians Gadgil and Guha, 'we expect the occurrence of substantial social strife. In fact, the clash of two modes has invariably resulted in massive bursts of violent and sometimes genocidal conflict' (1993: 53). The violent encounter between the Nagas and the British – a process that colonial rulers described as 'pacification raids' to bring order in the foothills – can be best understood in these terms. Between 1835 and 1851 the British carried out ten punitive expeditions in the Naga Hills in response to the innumerable raids by Nagas on tea gardens. The need to protect British capitalist interests (expanding tea plantations) in the foothills and the successive failure of hard pacification and non-interventionist measures during the 1860s and 1870s led to the adoption in 1873 the Bengal Eastern Frontier Regulation, or commonly known today as Inner Line Regulation, which formally prohibited plains people from entering the Naga Hills, but in effect had far-reaching impacts in protecting the new resource regime (Das 2009; Kar 2009).

The policy in practice divided the hills from the foothills and the plains of Assam proper. The legislation produced a sharp distinction between hill men and plains people. The colonial government could now segregate the revenue plains settlement from the hills, and this legislation brought about the distinction between the hill and plains 'civilizations'. The political project of the Inner Line was to provide a legal and administrative map that divided savage from civilized, the wasteful from the productive. The inhabitants beyond the Inner Line, the 'Nagas,' became a people without history, located racialistically in a state of nature before history began.

The Inner Line soon transformed into an artificial frontier between the hills and plains as a jurisdictional rather than a territorial boundary.[4] It established the dual world of capital and pre-capital, administered and partially administered, settled and mobile agricultural space. The shifting cultivators, who had long maintained trade and reciprocal relations and shared flexible territorial boundaries with the plains people, were brought under a defined geographical order by defining a line that also defined time. As one travelled from Assam proper towards the hills it was a process of transferring into wildness and chaos. The Naga shifting cultivators were now circumscribed by a territory and space that was marginal when viewed from the territorial framework of the state. In Tsing's words (although describing a different area), it was 'a zone of not yet mapped, not yet regulated' (2008: 28).

As Kar (2009) observes, 'the Inner Line was thus not a line in territory; it was a line in time. The advance of the line on the map was read as the progress from pre-capital to capital, from time of "no law" to the time of "law" '. The Naga *jhummias* were now placed in a historical political consciousness of empire's margins and frontiers. Thus, the study of shifting cultivation I undertake in this book goes beyond the analysis of *jhum* per se. It is also a political project of understanding Naga–state relations as they have evolved over time as colonial administration began territorializing the Nagas through law and as the 'sociology of knowledge' on Naga tribes developed through written ethnographic monographs and visuals, documenting Naga customs and practices that established their identity as a 'primitive other'.

The Naga–State Relationship: Modernizing Agriculture

Since the mid-nineteenth century, as colonial survey parties began touring the Naga Hills, the Naga tribes were framed within a cultural lens as hill people, ruffian and savage. Detailed ethnographic accounts published through monographs (Hodson 1911; Hutton 1921a, b; Mills 1922, 1926) in the early twentieth century brought to light their customs and traditions, political systems, social and kinship structures, beliefs and aspects of material culture that became important for administrators. This textual description of Nagas and tropes of representation were also articulated through photographs and sketches of their bodies, appearance and their artefacts, which became characteristic of their image (Woodthorpe 1874; Mills 1922; Butler 1926; Fürer-Haimendorf 1937).

In order to effectually govern the Nagas, the colonial administration quickly realized that they should build patronage relations with village chiefs and tribal elders. Hard military interventions were ultimately counterproductive (see Chapter 3) and direct administration of the Hills was out of the question due to the slim prospects for revenue generation once permanent control was established in the late 1870s. Thus, the colonial administration worked with village elites by recognizing their leadership as heads of Naga 'village republics'. As tokens of recognition and goodwill, the British administrators gifted red shawls to village heads, known as *gaon burha* (village headman). As J. P. Mills writes:

> Ao Nagas have no hereditary chiefs, and the council of elders is too large and cumbersome a body to act as go-between the village and

> officials. Headmen are therefore selected for this purpose, a small village having one and a big village up to five or six. These are in practice chosen by the village and approved by the Sub Divisional Officer, the men being selected are of good position in the village and can state a case clearly. They can be dismissed for misconduct, but usually hold office till death or till they become too infirm to carry on. (Mills 1926: 407)

It is through the recognition of headmen as village authorities and the selection of able-bodied men (people who enjoyed authority and prestige), commonly known as *dobashis* (DBs), or political go-betweens, that colonial administration interfered and had dialogue with the local people. The term *dobashis* etymologically means a person who can speak two languages. These individuals played a key role in assisting the day-to-day work of the colonial administration. They were carefully selected from Naga society, individuals who wielded respect among their kinsman and had influence over their tribe and neighbours. Even agricultural improvements were promoted with the help of the village headmen and DBs who became the link between the administration and the village. This process was not unique to the Naga Hills in the sense that even in the peasant societies of mainland India, the office of the village headman was upheld as the prime channel of communication between local administrations. However, there are differences in which the Nagas attained special attention. Unlike in the rest of British India, in the Naga Hills the political exigency of the frontier made it necessary for the colonial administration to act with caution. Direct administration was not considered; rather the aim was to maintain indirect control through this network of political go-betweens and through diplomacy and skilled control of local elites and clan heads, first settlers and village people who were men of repute and honoured in the community. In the colonial unadministered parts of Naga Hills, the village headman and chief, as Elwin (1961: 28) writes, acted as a command *dobashi* or go-between in the British pacification raids over headhunters. They also became the local contacts, informants of raids and headhunting, collaborators and acted as messengers in these territories.

The village heads became the key instrument through which colonial administration could extend its control over the Naga villages, settle disputes and aid administration in pacification raids against headhunters and slave traders. In the administrated parts of the Naga Hills District, government was firmly established over the Mokokchung and Kohima subdivisions. In the early twentieth century, the colonial government started introducing experimental crops, often tested in the fields of village headmen. We came to know from villagers in our study area that the first agricultural demonstrators (*kelu babu* in Yimchunger language) came from the Angami area and lived in the village headman's long house. The science of settled agriculture that was demonstrated by these local Angami demonstrators was facilitated through colonial local administrators. This structure of governance was not simply a case of dominance, 'but also a distinct colonial political system linking the colonial state and indigenous elites together in a common political order' (Gilmartin 1994: 1128). The Inner Line set the limits of revenue settlements and state forest in the Assam–Naga Hills frontier. The Nagas were outside the colonial remit of registered census subjects – they were living in the fuzzy border zone as lightly administered subjects.

Colonial administration rather than colonial science played the most significant role in *jhum* improvement during this time. Even colonial *jhum* land legislation like the Chin Hills Regulation of 1899 and the Naga Hills *Jhum* Land Regulation 1946 mention very little about land use and conservation. This legislation established the rights and duties of the local elites (headmen) as go-betweens and local authorities in administration. Similarly, colonial photographs that constructed the self-identity of the Nagas as the 'other' were instrumental in shaping the identity of 'chiefs' and headmen among commoners who were the lineage heads, first settlers and owners (*maalik*) of Naga 'village republics'. In colonial photography, the chiefs got special coverage as heroes: brave men, village heads who acted as collaborators in empire's administration over the hills (see Image 1.2).

In India's post-independence period, the intermediaries proliferated as political parties emerged and democratic representation became significant alongside counter-insurgency. The new intermediaries were development brokers in the village – village-level workers and party workers who took up the role of agricultural demonstrators but were not only responsible for the proliferation of wet-terrace rice technologies but also acted as brokers in development work, and as political henchmen loyal to counter-insurgency government forces. The role of the village headman and his council of elders was formalized and democratized through village development legislation passed in the post-independence period. A new set of village-level armed forces known as the Village Guards was established in 1958; these were raised by Major Ralengnao Khathing MC, MBE (1912–1990) popularly known as Bob Khathing to contain the insurgents in the village and to act as the main informants for the government's paramilitary force. In this book, I will also look at the life narratives of my respondents who have become important village brokers in development programmes and act as intermediaries.

Image 1.2 Moyang, the highly popular Konyak chief.
Source: Fürer-Haimendorf Digital Archives Collection, SOAS.

Framing Slash and Burn as a Problem

The colonial concerns on shifting cultivation were carried forward in the post-colonial period through neo-Malthusianism that linked population pressure on the land with swidden farming and consequent deforestation. At the end of World War II in 1945, the Assam administration, as part of rural reconstruction, took up a special scheme called the 'Grow More Food' campaign (see Chapter 4). This was integrated into the community development programmes during the five-year plans. Since the 1950s, special provisions have been made to tackle shifting cultivation though community development programmes and the introduction of new agricultural technologies with the help of agricultural demonstrators. One of the major policies pursued since the 1950s in the Naga Hills was the intensification of wet-terrace rice cultivation under the

North Eastern Frontier Area administration in the Tuensang Area that later became part of the Naga Hills in 1963. In the 1970s, the Nagaland government initiated a similar programme of wet-rice popularization through crop subsidies, loans and credit supplies and agriculture demonstration projects.

By the 1990s, neo-liberal policies of markets and decentralized governance changed to develop upland swidden farming though agroforestry programmes had shifted the emphasis from solely wet-rice subsidies and intensification to a variety of plantation, horticulture, soil management and wasteland development programmes implemented through community-driven development. This led to the initiation of transnational cooperation based on valuing local knowledge of *jhummias* for sustainable agriculture. The policy focus has shifted from the abolition of *jhum* to building upon local knowledge of biodiversity to develop sustainable *jhum* farming (Karlsson 2011: 12). I focus on one such programme, NEPED (Nagaland Empowerment of People through Economic Development), which started in 1995 in Nagaland and was built around participatory community development for *jhum* land improvement. The change in emphasis from earlier periods has been to reverse planned centralized agricultural development by building upon communities' 'social capital' and 'indigenous knowledge' (NEPED 1999, 2007).

The NEPED project started with the idea of community participation in afforestation programmes and *jhum* development and the diffusion of locally sustainable land use fallow management practices and the Angami Naga alder-based *jhum* rotation, in which pollarding of alder trees was central to sustainable crop management. As the project developed into its second phase in 2001, it broadened its focus from afforestation to incorporate gender empowerment, using micro-credits and creating self-help groups for the socioeconomic development of *jhum* cultivators. However, the project, like the colonial discourse, has framed swidden cultivators in Nagaland as egalitarian, self-subsistent, non-enterprising, and isolated – which denies the long-standing interaction between semi-nomadic agriculture and external communities. The universal vision of poor slash and burn farmers within such projects also overlooks social differentiation, the layer of power relations played out at the local level in the study village, Leangkonger (see Chapter 5). As Tania Murray Li observes:

> Upland swidden cultivators have been viewed variously as innocent tribes maintaining distinctive tribal ways of life, as peasant farmers, albeit perhaps as inefficient ones, as destroyers of the forest, and illegal squatters, and more recently, as expert environmentalists, holding the secret to sustainable and equitable community based resource management systems. These widely divergent perceptions indicate that there

> are potentially conflicting interests at work in the contemporary swidden landscape of the hills. (1999: xiv)

Agroforestry programmes as adopted by the NEPED project have gained popularity because they seem to provide a singular solution to the problem of upland marginality. Li observes:

> They satisfy many actors in the state government, NGOs, social activists as they provide a single script for action. However missing from this picture is up-landers interest in, and vulnerability to process of, accumulation, class formation and social differentiation one that can happen even in presumably highly egalitarian social order marked by non-class relationships of power. (Li 1996: 30)

Several case studies have shown how the transition from swidden to agroforestry systems has led to capture and the expansion of agriculture by elites in the name of afforestation (Suryanata 1996). Thus, we need to be aware of what Li describes as 'strategic simplification' (Li 1996: 31) embedded in agroforestry programmes.

In Nagaland, the introduction of community development programmes in the 1990s – with decentralized planning through village development boards that began in the 1980s – and the rapid march of state-led development projects in people's lives – have connected the local with the global. Swidden cultivators, however, are often more concerned over the control of their own resources. As one of the NEPED field staff who became my close friend during fieldwork explains, 'people are less bothered about conservation; they are more interested in immediate returns from protection of wildlife and biodiversity from poachers'. This becomes clearer as I subsequently attended a NEPED workshop during my stay in the study villages. This is where villagers participated in a daylong meeting to understand the importance of biodiversity conservation promoted by NEPED project staff. Participating villagers positioned their needs and arguments for conservation on their individual and community needs. Some community leaders proposed the idea of creating a joint village-level wildlife conservation committee that would impose fines on poachers, while others offered suggestions to constitute an honorarium for forest guards who would protect village forests. My host proposed his grand scheme of collaboration with neighbouring villages and implemented the mandate of the existing village-level wildlife conservation committee, of which he was the president. The various responses from village leaders representing their community made me aware that questions of how the environment should be managed became secondary to villagers' self-interest; issues like who gets to manage them and whose interests are served become central in local discussion.

Importance of the Book

The book has several points of interest that are insightful and unique from what has been written on swidden farming and forest history in South Asia. It analyses the political economy of *jhum* farming in Northeast India from a social ecology perspective and the interaction between space, culture and resource. It opens up the discussion that the post colonial frontier of Northeast India does not present a homogeneous place, but is mediated by the practice of colonial territorialization, based on the politics of 'cultural difference' and the construction of the other. The intellectual contribution in this study goes beyond the realm of Northeast India – the analysis is of relevance to swidden studies across the globe.

First, I try to explore the cultural difference attached to swidden cultivation (practiced in the hills) and lowland wet-rice farming, not as mere 'ecological', 'geographical' or 'location' domains but on how the relationship between the highlands and lowlands throughout colonial and post-independence period has been socially constructed and politically mediated. To put it differently: in order to link the conventional concerns of ecology with those of political ecology, marginality and history, I explore the swidden cultivation landscape as a component in the national and global systems of meaning, power and production (Li 1996; Ellen 1999).

Second, the book will present a non-normative and non-instrumental analysis of the changes that have come in the swidden cultivation landscape beyond the neo-Malthusian approach of population pressure and environmental change. The argument in the book is geared towards understanding agrarian change as an ongoing dialogical process shaped by the Naga tribes' encounters with the outside world (with missionaries, colonial and post-independent states and markets). The coming of colonial administration, missionary faith and English education all had overlapping effects on the Naga way of life and their vision of progress and improvement.

Third, the importance of the book lies in studying the Naga social system and communities in transition through historical ethnography and by engaging with multiple sources based on oral accounts, visuals (photographs), official accounts, colonial ethnography, missionary accounts and tour reports and diaries of officials on duty. These multiple perspectives are important 'to read against the grain' (Uberoi et al. 2007: 23) of existing ethnographic work on the Nagas and to draw upon the 'social construction' of Naga life by what Moore and Vaughan called reading insiders' and outsiders' accounts and placing them in perspective, beyond the normal anthropological descriptions (Moore and Vaughan 1994: 8).

Finally, the book spells out the idea of *jhum* development, drawing on the narratives and life histories of people engaged in the NEPED project and village intermediaries, through what Agrawal and Sivaramakrishnan (2000) have called the 'story telling of the development process'. I want to consider the multiple meanings and connections that adhere to shifting cultivations practice, as well as work towards an alternative framing of environmental concerns outside the causative frame of 'problem and solution' discussed in past work on shifting cultivation.

Structure of the Book

This book is based on historical data, ethnographic evidence, oral interviews and life histories. In order to avoid potential complications in the structure and design of the chapters, and to preserve the clarity and focus of arguments, the chapters have been organized thematically. The work is an interdisciplinary effort blending political anthropology, development anthropology literature, colonial photography, environmental anthropology and social history. The work begins in the late nineteenth century with the British involvement in the region and ends in the current period to trace the impact that missionary modernity and British administration, viewed as a regionally specific set of processes unfolding in Naga Hills, had on Yimchunger Naga agriculture, land use, food habits and identities. The book has a historical focus, combining both the colonial and post-independence (contemporary) phase of agroecological development in the Naga Hills. By focusing on Naga swidden cultivators, the book tries to achieve two things. First, it aims to highlight the agroecology of the Yimchunger Naga tribe that has not been part of the colonial ethnographic exercise of detailed documentation through monographs, and secondly it aims to understand agrarian change from a community level village-based study. It is not limited to a micro-level case study of a community swidden farming practice, but rather it connects to wider themes and historical processes (colonial and missionary encounters) that have shaped swiddeners' marginality and belonging to a place – the hills. The book is divided into six core chapters along with an introduction and conclusion.

Chapter 1 presents a general introduction to contemporary debates on shifting cultivation and the political context of Naga–state relations from colonial encounters to contemporary times. This will allow readers to grip the political and social context of the study area and its people.

Chapter 2 examines the fieldwork encounter and engagement with ethical and moral dilemmas while undertaking field research in a region that has been the centre of intense political activity between the state and the Naga groups. The chapter will provide an analysis of the dilemmas of fieldwork that raise epistemological questions of putting oneself at risk of harm for the sake of fieldwork.

Chapters 3 and 4 look at the history of the Naga's relations with the colonial state and the post-independence Northeast Frontier Area administration. I will discuss how colonial state control was exercised through violent raids and slave pacification tours that led to the extension of the colonial state and its control over the unadministered tracts bordering Burma's frontier. Parallel to these military expeditions, I trace how the colonial administration carried out improvement work through the promotion of new agricultural technologies that were aimed to settle swidden farmers. The policy pursued was one by which colonial administrators selected local wet-rice cultivators to train the Naga *jhummias* in this crop's cultivation. Aside from this, ideas of conservation were obtained from what we know today as indigenous knowledge systems of fallow management (the alder-based *jhum* fallow) practices by Angami Nagas to control the nomadic circulation of *jhum* plots. Colonial legislation could not prevent swidden cultivators from using their land; rather the exigencies of the frontier compelled the hill administration to work closely with village heads and local farmers towards spreading ideas of settled agriculture. In this process, documenting local customs and practices became important for successful administration. It also gave rise to an ecological consciousness of *jhummias* as 'forest destroyers' and wet-paddy rice farmers as 'forest guardians'. The depreciatory stigma attached to *jhum* as 'primitive' and 'backward' has found self-identification among the eastern Nagas of present-day Tuensang, Mon and Longling district as they define themselves as unadministered Nagas and hence backward people.

In the second part of the book, I sketch the narratives of agrarian change by introducing the Yimchunger village and engage in village-level ethnography to understand land and labour relations in a Naga village. Aside from this I will document changes that have come to farming practices. In Chapter 5, I particularly pay attention to land and labour relations that define agrarian change and how new forms of settled agricultural practice such as plantation farming and agroforestry help establish permanent tenure and generate new regimes of access and control over communally owned swidden plots.

In Chapter 6, I look at the missionary discourse on time management in the Yimchunger Naga village and how labour relations are regulated by the

Baptist Church followers. The chapter presents important insights into changing land and labour relations in Naga villages. It also examines changing food habits, which induce and inform the cultivation of *jhum* crops like millet to terrace crops such as paddy. The chapter engages with the agency of the community and the reformation of identity and morality that has far-reaching impact on traditional lifeways, divinity and social ethics as has been identified in other anthropological studies on modernization and globalization in post-colonial societies (Ngo 2016: 5; Hafner 1993: 3). Baptism is instrumental in bringing about a change in lifestyle, food habits, work ethics and social values influenced by puritan dogma. Nonetheless, missionaries worked with local customs and a tradition of labour organization in the village and demonstrated how 'age sets' become central to the missionary discourse in bringing about improvement through a shift in choice of crop cultivation from millet to rice. The chapter also exposes the syncretic cultivation of Baptist faith and tests the commonly held view that missionary influence only negatively influences the Naga lifeworld through modernization.

In Chapter 7, I examine the NEPED project. I discuss how Naga communities that have become beneficiaries receive the project subsidies and development packages for sustainable community-driven programmes. Concurrently, I show how published NEPED project literature such as annual reports and bulletins reflect on the success and failure of the development interventions to improve the livelihoods of Naga swidden farmers. In this activity, project staff project a new identity onto the swiddeners, which takes us back to colonial stereotypes and ideas of modernity and improvement.

In the concluding chapter, I analyse the changing swidden landscape in Northeast India and what possibilities it holds for swidden studies in South Asia. The book is not only about shifting cultivation per se. It engages with a wider socio-political milieu in which swiddeners – *jhum* cultivators – interact, collaborate, resist and cooperate with other actors outside their daily toil of cultivation (project staff, engineers, government officials and missionaries among others). It is also an ethnography of village elites, paddy cultivators, missionaries and state/project officials who have come to define the world of swidden cultivators. In this marginal world of shifting cultivators there is a creative process of resistance, negotiation and aspiration for improvement. Anthropologists have long been preoccupied with documenting cultures dying and disappearing through 'salvage ethnographic documentation', under the onslaught of missionary modernization[5] and Western influence. In this process, there has also been a misrepresentation of the swidden cultivators as non-changers and of swidden agriculture as a dying tradition.

The book is an interdisciplinary effort to revisit the pejorative association of slash and burn agriculture with ecological destruction, doing so by outlining the changes that have come in swidden cultivation landscape, leading to the rural differentiation and globalism of local community life. To this end, the book proposes to ignite critical reflection on *jhum* from an 'emic' perceptive and peoples' shared experiences of agroecological and land-use change in highland South Asia.

2

Methodology and Fieldwork: Negotiating Hazardous Fields

> In many areas of the world, anthropological fieldwork is more dangerous today than it was in the past. There are approximately 120 'armed conflicts' [...] and given that about one-third of the world's countries are currently involved in warfare and about two-thirds of the countries resort to human rights abuses as normal aspects of their political process to control their population, it is clear that few anthropologists will be able to able to avoid conflicting situations and instances of socio political violence in the course of their professional lives. (Sluka 1995: 276)

> The methodology underlying anthropological fieldwork has undergone great changes over time. In the days of Evans Prichard, Malinowski, et al. it involved the study of 'primitives': non-Western culture by anthropologists from a Western perspective. There is hardly any society now, which is 'closed'. Education has spread far and wide, and people in the so called 'primitive societies' now respond to and are aware of outside influence. Fieldworkers of my generation have to be aware of this situation. (Brara 1998: vi)

The chapter will discuss the moral and ethical dilemmas of conducting fieldwork in a social context in which everyday violence and suspicion features predominantly. In this context, methodological practice such as reflectivity, informed consent, fieldworker's responsibility, rapport and reciprocity need to be considered to engage with the ethics of fieldwork. In ethnographic (qualitative) fieldwork, the question of a researcher's subject position becomes fundamental as he or she is expected to engage with and become immersed in the situation and, at the same time, be objective and neutral. This becomes even more difficult in dangerous field sites where the researcher's wellbeing is constantly at risk. The fieldworker is mandated to generate 'thick description' (Geertz 1973), and in this process is obliged to spend time with his respondents, friends, interlocutors–informants in a way that goes beyond conventional data collection in qualitative research practice. In the process of recording data, the fieldworker encounters anxiety, dilemmas and undeclared challenges. A range of ethical issues emerges, which that I will discuss in this chapter, as one creates a set goals for fieldwork that cannot be considered during the planning and permission stage.

Simultaneously there is another ethical dilemma that I encountered while doing fieldwork in Nagaland. While using survey, interview data collection and ethnographic methods in the same context, I sometimes found myself in a very controlling position (recruiting subjects, distributing surveys, conducting interviews at a specific time and place). And sometimes I occupied the very opposite data-collection position, with little or no control over the research context, while attempting to blend in and be as unobtrusive as possible in order to better explore local routines, procedures and assumptions. I did not find these rapid shifts in subject position of research easy, especially in a risky fieldwork context. In unfamiliar settings that carry an element of threat, the subject position of the controlling researcher seems far more secured and reassuring. I am inspired to pen this chapter in minute ethnographic detail, because I believe it is time that anthropologists engage with the ethical quandary of putting oneself in harm's way in dangerous fieldwork settings. I hope my ethnographic vignette will help future generations of ethnographers learn from my experience, and that anthropology associations will reframe existing research ethics guidelines on working in dangerous environments.

The Violent Encounter

Let me begin my narrative with the violent encounter. It was on 11 June 2009, after spending 12 months in the field, that I was exposed to a violent event that remains an important mark of my fieldwork. After arriving in my study village, Leangkonger, in Tuensang District (the village name is spelled in the government records as 'Leangkonger' while Yimchungers call it Leangkonger), I settled with Nikon's family who were politically influential in the village. My host's nephew and his sons assisted me with my household survey, which focused on livelihoods, household consumption patterns, land use and land relations. I used questionnaires to carry out a baseline household survey in the village as part of my fieldwork. In this household survey I covered 126 households out of 140.

My study was not about violence (at least not in the conventional sense), but was focused on the effects of development projects. The road connecting Leangkonger to the district headquarters (Tuensang town) was maintained by local armed law-enforcement contractors. Although it was risky to travel on such mountain roads during the peak monsoon season, with boulders and mudslides cutting off the vital road links, I had planned to visit Shillong, the capital town of Meghalaya, my home town, and intended to return to my field site on 11 June (see also Chapter 5). Also, my host from Leangkonger was in Kohima to visit his political party and seek support for development work for the village and his farm. I had decided to meet up with my host at Dimapur on my return journey and then travel with him back to Tuensang.

On 9 May I reached Dimapur. The same day we travelled through the rugged mountain highways of Kohima and Mokokchung, then onward to the plantation of my companion's leader near Shamator subdivision headquarters in Tuensang district. Before leaving, my host and his leader's parents prayed for our wellbeing, holding the Bible in their hands. Every journey into the wilderness of Tuensang district entailed not only the danger of landslides but also the threat of being ambushed by insurgents and stopped by the Assam Rifles.

As we reached Tuensang town after much struggle through the monsoon rains, it was already dark. Leangkonger still lay 73 kilometres away. That morning, when I had reached Dimapur, the driver informed me that his boss had left; this turned out to be a miscommunication but perhaps intentional. I later learned in the field that our journey had been kept secret, known only to ourselves and the host family. The bumpy road had already made me drowsy: after travelling for over 270 kilometres and ten hours, I was falling asleep. As we were running out of fuel, we had an urgent need to refuel in Tuensang town, but as it was Sunday, the sole petrol pump was closed. Earlier, on our way from Mokokchung, a drunken man had suddenly appeared on the road, waving at us to stop. Our driver averted an accident by driving past him, accelerating sharply. Now, as our vehicle left the petrol station, it was swerving around corners. Half asleep, suddenly I saw our car being stopped by two unidentified armed men. The one on the left was carrying a revolver and the one man on the right was carrying an assault rifle, pointed at the vehicle. The vehicle immediately stopped with a violent jerk that threw us from our seats. Within minutes one of the armed men had taken control of the vehicle. They had already smashed the headlight, so everything was happening in darkness. Before I could say a word, our driver was bleeding from his forehead where he had been smashed with a pistol butt. My host, who was trying to intervene, was dragged out of the front seat. In the meantime, a group of armed guards entered the vehicle from the rear and ransacked my belongings. The militants were shouting, 'Don't you respect us as soldiers? We are in civilian dress, you don't respect us. You will have to pay for it with your blood.' The driver pleaded for mercy. In the meantime, the assailants were holding my host by his collar: he already had developed a swollen forehead. I remained untouched at that point.

In the midst of this violence, I heard an armed guard speak to me from behind, 'run, run, before you are beaten up too. Run for your life.' I called to the driver and we started running in the darkness with a dim torch in my hand that I had fortunately taken out of my bag before the event. After running for around ten minutes along the dark road, I called a college teacher who lived nearby and whom I had contacted earlier when the vehicle was being refuelled. We rang the police but to no avail. Soon I received a frantic call from my host. He tried to convince me that the situation was now settled, and we would soon be on our way to Leangkonger, so long as we came back to the spot where the incident had happened, or else the situation could get worse. I immediately informed his children in Tuensang town and they came to the teacher's hut, shocked and horrified. As the driver and I, together with the host's children, rushed to the site of the incident, most of the townspeople left, including the college teacher, who implored me not to go, as the situation was too dangerous. But I had little choice. It would have been morally incorrect to leave my host alone with the assailants when he was calling for help. The driver was very reluctant to come, as he, too, was fearful of being killed, but I convinced him. Moving through the darkness towards the scene of the incident, we realized that armed guards had cordoned the road. But soon I saw my host coming down the road with a loin cloth across his face, drenched in blood. He identified us as his associates. The next moment we were taken captive and escorted to an unidentified place, separated from the host's children. Threatened, and ordered to disperse from the spot, the townspeople politely agreed to leave without confrontation.

I had no idea where my field notes, camera and laptop were. They had been thrown out of the vehicle. All mobile phones were snatched from us and switched off. We were suddenly disconnected from the outside world. The dialogue that followed was emotionally charged. The town commander who stopped the car at times exploded, narrating his hard life as a rebel leader hounded by the military. Pointing at me, he denounced me as an Indian: 'Your government has repressed our aspiration for self-rule.' He also accused my host of misunderstanding them. 'The Naga army has been cursed to be *musafir* [wanderers] in the forest and jungle of Saramati mountain.' But, by this time, the paramilitary forces had been tipped off by one of the town headman and they reached the spot and rescued all the captives. The police and the assailants had come to a compromise for our release. In the days that followed, there were intense negotiations between the assailants and my host, who was an important political figure in the area. Scores of villagers, party workers and town sympathizers came to see him and express their support and concern.

Party workers argued that the assailants should be punished by the tribal council, invoking the tribal council rules, which contained specific provisions to punish assailants who harm civilians from their own tribe. However, my host's party leader was adamant that a situation of conflict should be avoided and proposed a compromise, which he announced, holding the Bible, as a 'compromise of forgiveness' as 'Jesus did to his enemies'. His stance was assailed by party colleagues and town sympathizers as an act of cowardice, conceding to the power of the guilty.

During these days, several tribal council meetings were held in the town. People accused the public leader of bowing down before the pressure of the puny underground town commander. The punishment imposed by the tribal council on the perpetrators had not yet been enforced. My host replied, bringing out the Bible from his bag, 'I am doing it in the name of the holy God'. He seemed convincing enough, but this moral stance also had a political purpose in deflecting the many people who said, 'he is our next MLA [Member of the Legislative Assembly], how can he bow down to such people who depend on him and beg him for money when they are in need?' Yet his position was a clever act of disguise. The retreat would hold the community together. My camera, notes and the vehicle were later recovered by the police and the Assam Rifles men at daybreak from the spot where our vehicle was ambushed.

For my host's party leader, this was a position he had to take in order to secure his political future. A lot was at stake: his farm, the whole political constituency and a wide electorate whom he could not possibly oppose or see fractured by antagonized factions. In our later conversations, I came to realize the importance of this compromise. The calm we maintained during the hostage crisis not only minimized the dangers of confrontation but also saved our lives. A compromise letter was signed between the two parties for reconciliation (see Image 2.1).

Immediate Reflections

Reflecting on the event, there were a few inferences that I could draw from it. In the immediate aftermath, I was more preoccupied by ethical and moral questions. One of my host's children asked me why I had initially run away when his father was beaten up by the town commander. I had no simple answer. In the days following my host and I again became good friends, but the question still haunted me: Had I betrayed him? The intensity of the violence had demanded quick action. However, many things were unclear. Why did they take their own community member hostage when they constantly stated, 'We Nagas are one'? What were the issues of conflict that led to this

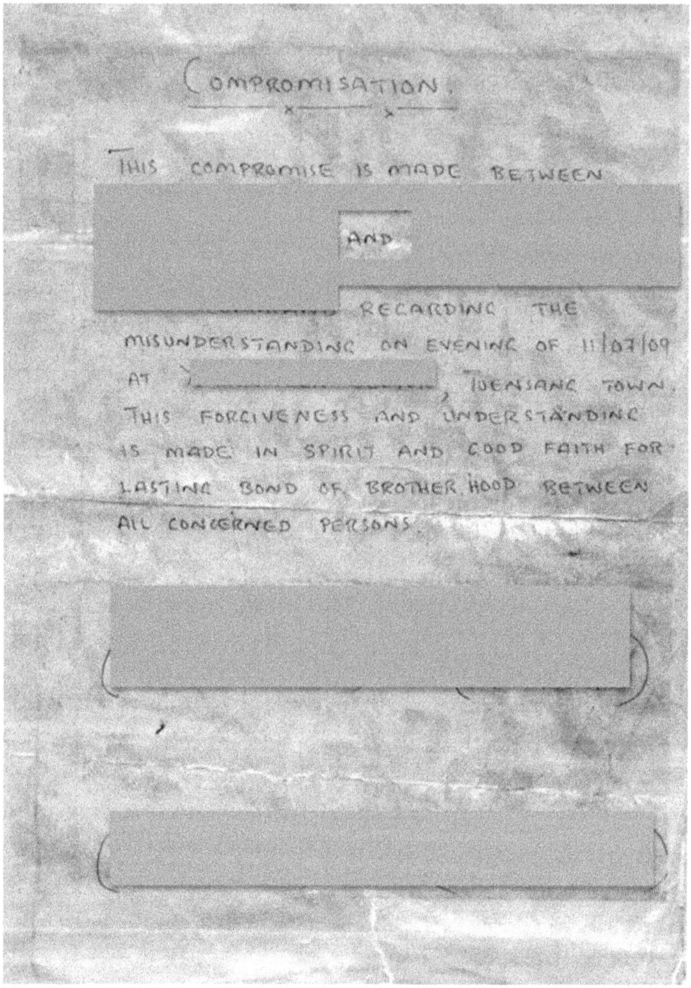

COMPROMISATION.

THIS COMPROMISE IS MADE BETWEEN
[redacted]
AND
[redacted]
REGARDING THE
MISUNDERSTANDING ON EVENING OF 11/07/09
AT [redacted], TUENSANG TOWN.
THIS FORGIVENESS AND UNDERSTANDING
IS MADE IN SPIRIT AND GOOD FAITH FOR
LASTING BOND OF BROTHER HOOD BETWEEN
ALL CONCERNED PERSONS.

[redacted]

[redacted]

Image 2.1 The Compromise Document.
Source: Author's photograph, 2009.

attack? Was the ambush pre-planned? What were their motives? While we were being held, we were listening to the town commander's narrative of his struggles, but the situation was too dangerous to have any meaningful dialogue with the assailants. As there was no possibility of taking down the exchanges between us and the town commander, I felt a lost opportunity to record an interesting dialogue in captivity. Reflecting afterward on what had happened that night, I felt I had no control over the production of ethnography. As Nilan (2002) has emphasized, in fieldwork the question of control over the recording and production of data causes much anxiety for the qualitative ethnographic fieldworker in an 'exceptional setting', where emotional detachment, objectivity and control over the production of data are difficult to achieve. In the violent encounter that night, the government forces did not apprehend the culprits – rather they acted as a go-between. Interestingly, nothing appeared in the local newspaper about this event. Every piece of information was concealed from the outside world. It helped in building confidentiality, as I also kept the incident confidential and did not share it with anybody. I thus acted as a good ethnographer during the entire research exercise, strictly following my research protocol and ethical guidelines. Only after returning from the field did I share the experience with the members of my research committee, who expressed their empathy and worries about conducting fieldwork in such a precarious setting.

After the Event: Distortion or Enlightenment?

Nilan has written that, although, when working in unfamiliar and dangerous fields, it is tempting to 'identify the most emotionally intense moments in fieldwork research as the pre-eminent moments of data collection[; in fact] it is often impossible to collect really informative and useful data at such times', emphasizing, rather, that 'it is often data collected after the event or in quieter moments which turn out to be the most evocative, in the end' (Nilan 2002: 383). In some ways, this was also my own experience. Like her, I found that it 'was really everything I found out afterward about the events which produced the most satisfying explanatory account of the phenomenon' (Nilan 2002: 383).

In the days that followed, I was determined to go back to Leangkonger, despite my fears, to continue my research on farming. My field notes and laptop had been recovered by the police. I was encouraged by the support extended by the villagers, who told me that the compromise letter would act as a safety valve, ensuring my security. Over the years, the village had become an important site of political activity. Roads, jobs and contracts came to the villagers just before elections. Political intermediaries in the village selected the individuals who would be beneficiaries of government programmes. The lives of the villagers were dependent on development schemes and projects. Since 2002, the passing of Nagaland Communitization of Essential Services Act, all state development programmes and projects had been administered at the local level.

My post-incident fieldwork revealed the patronage nexus that the villagers had developed with project officials, and that project officials in turn maintained a relationship not only with political parties in power but with underground groups. When the interests of rural elites and opposition political parties clashed, violence was the inevitable outcome. However, I was to discover that the role of underground factions in this scene went beyond the violence of hostage-taking. It was embedded in the process of accounting for development monies.

My decision to return to fieldwork in the village after the hostage incident was thus crucial to my understanding. In the end, it turned out to be the most productive period. While initially the 'hostage crises' – an eruption of violence – seemed obviously not linked to my immediate research activity, I came to see how it was embedded in the day-to-day violence produced in these borderlands. Also, my perception of the relation between the violence of underground factions and development became completely inverted. When I first started my project, I was, of course, aware that the ceasefires and transnational development aid were intimately linked. Without the ceasefire, there would have been no aid, and equally, the aid was presented as bringing peace through development to an area understood since colonial times as an 'underdeveloped', 'backward' and 'rebel-infested' frontier state.

Violence and Its Interpretation

In recent years there has been an upsurge of anthropological studies addressing the question of violence. However, as Ghassem-Fachandi explains, the central concern has been with violent situations as the object of study, rather than as methodological and ethical challenges for the researcher; in fact, 'they usually avoid incorporating into analyses their own personal encounters with violence in the field' (Ghassem-Fachandi 2009: 4). However, there has been increasingly rigorous introspection on the methodological and philosophical concerns of fieldwork in unstable places. The issue first appeared in an unpublished, but much discussed, 1986 paper, 'Occupational Safety and Health in Anthropology' in which Nancy Howell argued for the need to discuss the issue of personal dangers involved in doing research. Howell's appeal for a practical piece of research that would produce a comprehensive survey of occupational safety and health hazards was quickly accepted by the American Anthropological Association (AAA), and an expanded version of her paper was published in 1990 as an AAA special report titled, 'Surviving Fieldwork'. Evaluating over 80 separate dangers, the paper identifies dangerous ethnographic encounters in the context of sunburn, robbery, venereal disease, stinging ants, frostbite, arrest, malaria and military attack. It provided some compelling data on aspects of human hazards that were very difficult to quantify – suspicion of espionage, acute conflict and political turmoil (Howell 1990).

Since then the debate on fieldwork dilemmas in hazardous and perilous environments has developed far beyond occupational health and safety. Kovats-Bernat explains the expansion of the issues recognized:

> As we continue to conduct research in increasingly hostile and dangerous regions, the very real possibility of our victimization in the field presents a challenge not just to the practicalities of personal safety but also to the ethnographic methods and ethics that we are retrofitting for use in cultures in which ordinary interrelations and social institutions are overshadowed by unrest, instability, and fear. Even exposure to low-intensity repression or harassment over the course of research threatens to adversely affect the ways in which we approach the field and interpret social phenomena within it. (2002: 208)

Thus, the further problem that arises is epistemological, a methodological concern with 'how the status of field hazards affects the ethnographer's observations and data analysis' (Kovats-Bernat 2002: 208). Violent encounters not only threaten the ethnographer's physical integrity but also potentially damage the integrity of their enquiry. As Feldman puts it:

> Violence never stands in a relation of pure externality to the consciousness of the researcher. One experiences it through the moment of crisis and strife and internalizes it and expresses it afterwards from memory. The inability to extricate oneself from what one studies is fundamental in such experience, where the moral and physical integrity of the ethnographer is put at risk. Yet, at the same time, such extraction remains a necessary moment in order to arrive at some degree of objectification. (1995: 228)

Further, in witnessing violence, the ethnographer becomes a part of what is being studied and therefore complicit in the further unfolding of the event.

Against such negative effects on understanding, it has also been argued that 'the subjective experience of violence brings about, or allows for, an insight about the realities in a particular place' (Ghassem-Fachandi 2009: 4–5). This experience made me aware of the political economy and the rationale of violence in the Naga Hills and is closely linked to the way the state and civil society have historically been placed in opposition to each other, with intense friction, conflict, domination and subjugation of local aspirations.

At the same time, I became very aware of how violence is experienced by people who have previous experience as perpetrators but who now have become key narrators and interpreters of such encounters to the ethnographer. As Ghassem-Fachandi writes:

> The subjective experience of violence is peculiar in that it tends to overwhelm other daily experience. Violence humiliates, the victim is humiliated by being reduced to an object; the witness by being reduced to impotent bystander and the perpetrator, in retrospect, by an inability to ever adequately redress the violence perpetrated. Or, in turn, all the participants are humiliated because it is frequently difficult to delineate clearly between victims, bystanders, perpetrators or analysts. (2009: 5)

There is also a special difficulty in writing about violence, in 'objectifying that which has been consumed, harmed, or destroyed, that which is no longer there except in absences, traces, wounds and ruins' (Ghassem-Fachandi 2009: 5, describing Borneman's essay in the edited collection). The violence experienced can often lead to powerful textual expressions of experience – narratives that at times verge on a 'cathartic babel' (Ghassem-Fachandi 2009: 6), reflecting multiple points of interpretation.

The questions that surfaced with reflection on my fieldwork are: How do anthropologists cope with the narrowly defined research ethics in 'abnormal' dangerous field sites? Can we simply 'retrofit' (as Kovats-Bernat puts it above) existing methods and ethics? What are the moral dilemmas that confront anthropological research in such uncertain fields? And: What kind of methodological and ethical issues emerge from the practice of research collecting fieldwork data in chaotic and dangerous situations? My position on ethics centres on the fieldworker who enters the field for his ethnographic work. The ethics of putting oneself in harm's way while undertaking fieldwork in dangerous settings has quite different implications than the conventionally understood professional ethics of protecting the interests and welfare of interviewees or the study community. I will proceed with these questions and explore the ethical and epistemological risks that anthropological research encounters.

Ghassem-Fachandi observes: 'Rarely are anthropologists as field researchers prepared for the violent event that unfolds around them, much less for the ethical effects they have on them afterwards' (2005: 5). While doing coursework in anthropology, I was introduced to research ethics. The research methodology course was tailored to train students on ethical issues, research methods and methodological guidelines to conduct fieldwork. The guidelines and frameworks of fieldwork were quite silent on questions of violence, insecurity and the practical day-to-day encounters of researchers with their subjects in insecure field situations. It was often presumed to be 'common sense' to choose research areas and topics that would not invite danger. In my own case, I had not apprehended the severity of my fieldwork situation, as I was convinced that life would be 'normal' in the decade-long ceasefire zone.[1] I embarked on fieldwork in June 2008 after an initial short visit to Nagaland in 2007. While I was overwhelmed by the dangers I experienced during this visit, it seemed to me more of an adventure.

The Issues of Publishing and of Taking a Stance

As a pure scientific observer, I was taught to be 'emotionless', 'unbiased', 'unsympathetic' and committed to objective scientific enquiry. But the practice of detached observation obscured and made my ethnography less reflective. The temptation to become an activist was always present in me. I thought of reporting the violent incident to the press, organizing villagers to file a right to information (RTI) petition against the district officials and the man involved in siphoning wages from beneficiaries (see also Chapter 5).

As a good anthropologist, I decided not to report it to the press. I felt that I was least empathetic to the situation by not writing on the crisis that people face by living in an area that is home to a rebel and factional tussle in which poor farmers paid the price. In the violent events that evening, a Hindustani mason lost a limb as he was shot in his leg minutes before we were stopped and assaulted. Incidentally, I later discovered from interaction with him that he was working in my host party leader's native village and was involved as a construction worker in the church.

Although I partially seem to fulfil the ethical requirements of my ethnography, by being neutral to the situation, the question of taking sides was palpable; I questioned my own conscience in the situation. My fears became real one day when I met a Khiamniungan tribe lawyer in Tuensang town. He had invited me to his subdivision, Thonoknyu, the most backward area, he claimed, in Tuensang. His main intention was to persuade me independently to report on the suffering and marginality of his tribe in relation to the advanced districts in the state. As Frances Henry notes from her fieldwork in Trinidad, 'This answer was not satisfactory to some people [and], in fact when I refused to write and expose the situation, one person accused me of being an CIA agent in disguise' (Henry 1966: 553). For my part, I was questioned by my Khiamniungan lawyer friend concerning the relevance of my research when I could not independently report on the facts of borderland communities, which were neglected by state and local elites. I made a sorry excuse that my focus of attention was the Yimchunger, and it was too late to conduct fresh fieldwork on the Khiamniungan. If I had reported on these issues, which were close to my heart, I would have faced moral sanctions returning back home as many past anthropologists have faced[2] in their research by entering conflict-torn areas, and nearly costing them their research (see particularly Bourgeois 1991).

Political neutrality opens up the dilemma of ethics and the risk of being naïve and emotionally detached from political questions and moral ethics. By not reporting the violent incident to the media, I was undermining the daily stories of violence that have become part of everyday life in this frontier district. I maintained a personal detachment from the inner workings of local politics and factionalism. Henry, in her study on Trinidadian factory unions, states: 'The situation of being caught between two rival factions is certainly not a new one for the anthropologist; but when this occurs in the political arena and involves the policies and legislative programme of the government of the country, these problems take on a new and sometimes threatening significance' (Henry 1966: 554). In my case study these conflicts were directly linked to state development programmes and grants that were connected to questions of the livelihoods of poor farmers and village daily wage earners, where money was siphoned off by bureaucrats, politicians, faith based actors and factional stakeholders. People who resisted irregularities in state development programmes, such as the Mahatma Gandhi National Rural Employment Guarantee (MGNREGS) Scheme, were targeted and victimized by the interest groups. By reporting these events I could have potentially sent alarm bells to the people who took me hostage and further endanger my stay in the field.

Working in a Context of Violence

I entered the field at a time of a decade-old ceasefire. During my fourteen months of fieldwork in different villages in Nagaland, both the military and the factional rebel groups were regular parts of the landscape. During my stay in the village, the local underground groups made two reconnaissance visits, each time sending demand notes to the village headmen to supply food grain to sustain the marching party. This was in turn followed by the appearance of the Assam Rifles *jawans* (soldiers) with their commanding officers, who visited the headman's house to gather information about the movements of the militants. I also recorded stories of violent clashes in the past between government forces and rebel groups. In 1993, the village was turned into a battleground as a fight broke out between government forces and rebels who had taken shelter there. The village headman's son took me around to show bullet holes in the electricity power pole that symbolized the violent encounter (refer to Image 2.2). Following that encounter, the rebel leaders, suspecting the village leaders of having leaked the information about their presence to the military, surrounded the elders and insulted them by exposing their

Image 2.2 AK-47 bullet mark on the village lamp post.
Source: Author's photograph, 2009.

bottoms – to which the elders replied that insulting them was hardly a way to attain freedom. The underground people then denounced the elders as government agents. Such exchanges were the stuff of gossip; the real scene brought the rumours and terror stories to life.

The village itself was on the brink of a split between the Baptist majority and a Catholic minority. For many, to convert to Roman Catholicism was considered a 'sin', undesirable in the true Baptist spirit. In fact, the first Catholic church, set up in the village a few years earlier, had to be dismantled under pressure from the public and the Baptist Association. Today the village is divided between one Catholic *khel* (ward) and three Baptist *khel*.[3] Conducting ethnography in such a sensitive environment demanded reflective awareness and sensitivity towards the different positions. In their private talks, the Baptist pastor would complain about the town parish, while the Catholic priest would highlight the lack of financial accountability of the Baptist associations. Doing fieldwork under such circumstances was difficult. I chose to live with the most powerful and respected families in the village. The choice, although it obviously carried certain disadvantages, was practical and made sense. I developed a good rapport with the village headman's son, who was a respected citizen of the village. On numerous occasions he made me understand the local complexities of land relations and farming. He accompanied me on many of my visits. While access to these households was easy, the locals would not talk much in front of the village headman's son or relatives. While there were no caste hierarchies as such, poor families found it difficult to narrate stories in front of village elders who were members of the village council. I could only gain access to their narratives when I started visiting these households independently. In this complex ethnic and political setting, negotiating access was mediated with persuasion and by building trust and rapport with the community.

My friendship with the president of the Leangkonger Village Citizen's Union, a body established by the Baptist Church in the 1970s, was crucial as he extended patronage with the village people from all ranks. I also attended the village Sunday services and participated in church programmes. In this way I was slowly exposed to the inner workings of the church, which generated trust among the community. After a couple of months, the Village Women's Leader – a church member – smilingly remarked one day, 'You have now become a member of our Yimchunger family'. My associations with powerful figures, important in local networks of patronage, was helpful in reducing my level of personal risk within the village, as I was always assured by these leaders that nobody would dare to harm me. However, as the village was internally polarized, I soon realized that alliance with any single faction would hinder my own work. My field co-workers were church deacons, the village headman's son, a former Naga National Council (NNC) worker and the president of the Village Citizens' Union. I also developed a friendship with members of the village Student Union. Association with these different people, and establishing correspondence with diverse pressure groups minimized the risk of hostility or ill feeling with any of the households in the village as well as provided balance and counteracted the potential for bias.

However, establishing rapport with villagers who had divergent interests and social identities involved maintaining a measure of balance, and also presented methodological challenges. In his famous essay, 'Behind Many Masks' (1972), Berreman describes the problems of conducting ethnographic research in a community of conflicting interests, religions, cultures and lifestyles: 'Even in small and isolated Sirkanda, social heterogeneity loomed as a major obstacle to my rapport and my understanding, just as it does to people's relationship to one another' (Berreman 1972: vi–vii). 'People's relationship to one another' was analysed by Goffman (1959) in terms of the 'means by which people seek to control the impressions others receive of them' (Berreman 1972: 146). Berreman argued that this social integrationist, specifically 'dramaturgical', approach also be applied to the relationship between ethnographer and 'subjects'; in other words, 'impression management' is of methodological as well as substantive relevance. How does the ethnographer 'convey a desired representation of oneself', especially in the difficult situation in which 'acceptance by one element of the community does not imply acceptance by the whole community and [...] frequently, in fact, precludes it'? (1972: xxii). The challenge I faced was to create and manage the 'impression' I made on my hosts and interviewees who were split on denominational and clan lines. Often, they feared my presence would send warnings to people, who would then misjudge my intentions.

Encounter and Suspicion: Establishing Fieldworkers' Identity

As an anthropologist studying agriculture, I was eyewitness to many cultural events in the field, attending festivals, living with farmers in their fields and with their families, participating in Sunday Bible services in the village church, visiting the village headman's house and, on several occasions, I was hosted by local politicians, church pastors, village headmen and tribal council leaders. People with whom I interacted were often welcoming, frequently suspicious and at times intimidating.

During my stay among the Yimchungers, I was often questioned about my purpose in being there and also about what good it might do for them. My answer has been that I wanted to write a PhD thesis about their agriculture and social life and that it probably would do them no practical good. The past history of outsiders creates a space into which the 'newcomer' is received, a background against which he or she must negotiate his, or his fieldwork's, identity. My position as an outsider/non-tribal/non-Western anthropologist raised various suspicions in different individuals, from police officers to villagers who became my friends and guides in the field. The police felt I was intruding into the protected spaces of the Nagas. The local people sometimes identified me as a journalist (possibly a spy, possibly a mouthpiece for their causes) and at other times as a trespasser collecting information about valuable medicinal species and crops that would lead to the stealing of local wealth. Even so, while people were pleased to share information with a person from a foreign university, there was also rumour-mongering about why a person would travel such long distance just to learn about them.

In its most violent form, the 'outsider identity' has been remembered in people's minds through the atrocities carried out by the British and Indian armed forces. While the British punitive expeditions to punish former trans-frontier Naga head-hunter and slave-takers were part of collective memory, the actions of the Indian Army, in many ways reminiscent of those of the British, were still living memories for many. My physical features, resembling Aryan (North Indian) traits, often disappointed villagers, who memorialized the dead and remembered the crimes committed in Naga villages by the Indian armed forces in the 1950s. The decades following India's independence saw violent resistance and deployment of the military to quell Naga nationalist sentiments. There was also a long history of misrepresentation to overcome. As one of my Naga respondents expressed it, 'outsiders come and misrepresent our culture, tradition and write whatever they like and wish to write'.[4] His claim seemed reasonable to me; indeed, most contemporary literature on the Nagas by Indian writers, including academics, have been based on secondary research and without much fieldwork.[5] The writings are often done from a nationalistic perspective. There is also the older heritage of colonial ethnographic work carried out by generations of administrator–anthropologists in the Naga Hills, such as J. H. Hutton and J. P. Mills, the most prolific among colonial civil servants,[6] whose ethnographies are today charged with misrepresenting Naga culture, customs and traditions. Abraham Lotha (2007: 7) not only documents their innumerable errors based on cultural presumptions about 'the savage other', but describes how, after a public declaration by the Lotha elders (*hoho*), many amendments were made in the latest edition of Mills's monograph, *The Lotha Nagas*, published by the Directorate of Art and Culture, Government of Nagaland. Similarly, other scholars have faced fines imposed by the community for misrepresenting local culture and customs. Some researchers merely had to provide a feast for the whole village, while others had to offer animals and sacrifices.[7] These worries also haunted my research.

In the village where I conducted my fieldwork, I was for a long time unaware of a more recent breach of trust. Two German anthropologists had earlier visited the village on a short trip to document a man who had been identified as a 'tiger man' for their photo-book project. 'Tiger man' spirit possession (lycanthropy) among the Yimchunger Naga had already interested foreign anthropologists in the region (e.g. see Hutton 1920) and had been widely publicized by Christoph von Fürer-Haimendorf, who in 1939 wrote about it for a London newspaper after his 1936–37 fieldwork in Naga Hills. When I interviewed the 'tiger man' and his family in my study village, they were very upset that they had never been sent a copy of the book by these German scholars. They also demanded that I should send them my photographs and not be a betrayer like the previous academic visitors. Such breaches of trust by past researchers cast my integrity in doubt.

In one encounter, a tour operator, this time from an Angami village in Kohima, explained to me that outsiders 'come and steal our "bio diversity" [... and] we should stop these bio-pirates. They seldom get permission from the village authorities before entering the village and do whatever they want'. He was referring to the new economic opportunity provided by bird watching[8] in the village nature conservation site that had become a contested resource of interest for tour operators and guides from the village and outside who were the obvious beneficiaries of the 'Green Village' project in Khonoma. A few years earlier, an Australian anthropologist had conducted his fieldwork in one of my initial field sites, Khonoma village. He had been hosted by a respected villager in Khonoma and had carried out a GPS (Global Positioning System) survey in the village, 1 to map its vegetation and resources (particularly land use), which the villagers did not like. When he left the village, people started complaining about his presence. There were rumours about the secrecy surrounding his fieldwork. Some of the respondents I interviewed claimed that he never shared his information. He never disclosed his research findings to the villagers, who had hosted him and given him all the logistical support for his anthropological and ecological research. The researcher had said that, as a matter of policy and ethics, he could not disclose his findings. The majority of the villagers (except the people who had been his hosts) denounced him for 'piracy' of local knowledge. However, he was supported under a transnational conservation and livelihood programme and enjoyed the good will of the village council.

In the decade of ceasefire, Naga customs and tradition are being reinvented by the documentation of Naga folklore and oral tradition and their portrayal in coffee-table books. The Ministry of Tourism (Government of India) and Department of Art and Culture (Nagaland Government) jointly host annual programmes in December to celebrate the diversity in Naga culture. The annual event, known as the 'Hornbill Festival', attracts international tourists as the government promotes its brand of cultural tourism celebrating 'unity and diversity' (see Images 2.3 and 2.4).

A few months before I arrived in Tuensang, a team of young German and Swiss researchers had been welcomed with offerings of Naga shawls in the Upper Konyak village bordering Burma. At the end of their stay, the villagers who had given the shawls demanded a huge payment of money for them. This disappointed the local tour operator and his clients. The research-cum-tour trip ended in disaster, and the German team of researchers had to return home without achieving their goal of documenting Naga culture.

This encounter makes explicit the interesting questions of reciprocity, gifts and exchange in play in the research encounter. In fact, this theme was already implicit in some of the encounters above: for example, the 'tiger man'. A sense of reciprocity is crucial in building rapport with interlocutors. Often the lack of reciprocity on the part of the researcher destroys goodwill and results in unsuccessful fieldwork. Thus reframed, the issues are not so much misrepresentation but theft or, jointly, misappropriation. Colonial ethnographers such as Fürer-Haimendorf enjoyed full liberty to capture images of 'colonial subjects' as nude, fragile bodies. Indeed, he then published his fieldwork images in the popular press, in a series of articles between 1937 and 1960 in the *Illustrated London News*.[9] However, the power dynamics have changed over the years, with

Image 2.3 A public servant being guarded during the Hornbill Festival, 2008. *Source*: Author's photograph.

Image 2.4 Journalist photographing folk dancers during the Hornbill Festival, 2008. *Source*: Author's photograph.

the very nature of anthropologists' relations with the field and the subject position of the ethnographer doing fieldwork now being questioned.

Evidently, some past ethnographers of Nagaland have taken a casual approach to the ethical requirements of anthropological research, posturing a superiority of expertise and posing as official 'trustees' of Naga culture. Today, in light of limited access to the field and the strict security checks imposed on fieldwork, researchers have resorted to fast-track, often piecemeal, visual ethnography that has produced shallow ethnographic detail and historical representations of Nagas through media such as under-researched glossy coffee-table books that are often unintelligible to the anthropological research fraternity.

Reflections on Fieldwork and Ethics

To many of my colleagues, friends and relatives, going to Nagaland appeared to be an act of bravery, adventure or foolishness. While this attitude was largely the result of the media-based impression of Naga people, I was more intrigued by what Naga friends and well-wishers told me about eastern Nagaland, the location of my study area. They saw it as a misadventure given my status as an outsider, someone who had come from London to do research on the 'colonial' unadministered Nagas. These were the 'wild spaces', where people still practiced *jhum* cultivation, wore little clothing and were less civilized. They advised me to conduct ethnography in a nearby village close to Kohima town from where I could commute easily. Van Schendel, who worked among the Mru tribes of the Chittagong Hill Tracts in Bangladesh, writes "it is not only westerners and Bengalis who have read the nakedness of the Mru as a sign of primitivity; Chittagong hill people belonging to other groups have done it too" (2002, 951). The parallels with the advanced Naga's views of their eastern brothers are striking.

Given its emotive reverberations, surviving one's fieldwork is the most important task in a perilous and intimidating setting. For this reason, fieldworkers working in violent field sites need to possess a sensibility about the field, a practical knowledge of the ground reality, what Nancy Scheper-Hughes calls 'anthropology with one's feet on the ground', or 'phenomenologically grounded anthropology' (1992: 4). Then, as Campbell writes, 'violence changes the experience of fieldwork and raises new concerns about ethics, research methods and the anthropologist's relationship with the fieldworkers' (2011: 9). This point acknowledges the chequered nature of ethical practice in anthropology and the changing relationship between anthropologists and their informants. It is simply not acceptable to ignore past unethical practice or to claim that our research does not raise ethical issues (perhaps our inability to see the wider picture leaves us blind or indifferent to important moral issues) or to assert the high moral ground and condemn those who also assert the high moral ground but do not accept our position and arguments. It would be dangerous and utterly false to think that there are definite answers to any of these ethical moral questions. We need to think about them in both practical and theoretical terms.

Over the last decades, there has been some rigorous introspection about the methodological and philosophical concerns of fieldwork in unstable places – from Nancy Scheper-Hughes (1992) to Antonius Robben and Caroline Nordstrom's examination of 'fieldwork under fire' (1995). What we must think about more critically, however, is how we deal with 'violence' in the field alongside ethical issues without undermining our obligation towards society. These may seem certainly value-loaded. Nonetheless, if anthropology as a discipline is to progress in the twenty-first century, it should address these issues. The current literature on fieldwork practice also demands fresh insights into violent and extraordinary field situations. However, I find it difficult to determine who can be seen as responsible for such violent encounters involving the anthropologist in person. This brings us to the question of a potential ethos of 'putting oneself to danger' that is rarely discussed in anthropological ethics, which is, though, more explicit on issues concerning the security and confidentiality of our informants.

My knowledge of violence and terror became clearer after the fearful encounter I narrated above. Extended fieldwork experience therefore becomes critical in producing a valuable qualitative data set substantive enough for research. Ethnographers working in dangerous field situations are pressed with the challenge of innovating new strategies for their wellbeing while at the same time continuing to identify and explain the unique social interrelations that arise amidst crisis and strife. In my own 14-month stay in Nagaland, I engaged with the problems and steered my way through them, at times adopting alternative routes and tools of rapport-building with my respondents that were quite unconventional in nature.

In the 1960s, anthropologists began to respond to the crisis in the discipline created by the post–World War II emergence of new nations from old colonies. Anthropologists increasingly breached research ethics, working as spies or being used as spies by the state in anthropological research (Nash 1976: 225). In the contemporary context, this has included anthropologists working at home (in their own countries) studying ethno-nationalism, subnationalism, and separatist movements in indigenous culture. These roles have been particularly challenging as they have put anthropologists to work with complex societies where they have to deal with politically explosive questions that require acumen and skills that are distinct from just studying 'other cultures'. Negotiating with these complexities helped me in successfully performing my fieldwork. Also, I would like to add that the everyday talk of violence was reflective of the politics of violence and helped me understand how violence was indistinguishable from the texture of people's day-to-day life. The end result is a reflective memory that has helped me to understand and, at least partly, to articulate the dangers and complexities involved in fieldwork under violent circumstances. In the next chapter I turn to the violent headhunting pacification tours that shaped colonial hill administrators relationship with the Naga swidden cultivators of the then Assam–Burma frontier.

3

Ethnography, Violence and Memory: Telling Violence in the Naga Hills

> The processes of state-building are multiple and complex. Different areas were integrated into states in markedly different ways at different times. (E. R. Wolf)

On 24 January 1938, Henry Balfour and J. H. Hutton were the discussants of a very interesting paper entitled 'Through the Unexplored Mountains of the Assam–Burma Border' and presented to the Royal Geographical Society by Christoph von Fürer-Haimendorf, who had returned from a year's fieldwork in the Naga Hills. The paper soon appeared in the Society's *Geographical Journal*. It described a two-week expedition carried out in the 'unadministered' and unmapped Assam territory of the Kalyo-Kengyo Nagas in 1936, where Fürer-Haimendorf had the privilege of accompanying the deputy commissioner, J. P. Mills, on what Balfour, introducing the paper, described as 'one of the unfortunately necessary punitive expeditions into a portion of the Hills never visited before by Europeans' (Balfour, in Balfour and Hutton 1938: 216)[1] (see Map 3.1). Known as the 'Pangsha Expedition,' its objective was to punish Pangsha Village, where villagers had been reported to have captured slaves from the Political Control Area (Chang territory) to trade into Burma; they were also accused of taking heads. The expedition was remarkable in many ways.

For Fürer-Haimendorf, the expedition was a remarkable opportunity to venture into an area where 'no survey party had ever penetrated' (Fürer-Haimendorf, 1938: 207) and to reveal the 'complete mystery [that] has […] veiled the Kalyo-Kengyos, of whom little more was known than their existence and the fact that they live in houses with slate roofs' (Fürer-Haimendorf, 1938: 202). En route, and on the return journey, the expedition met tribes that 'have been visited only occasionally by punitive expeditions and survey parties' (Fürer-Haimendorf, 1938: 202), including the Yimchunger, one of whose villages (Sanghpur) was also visited during the tour.

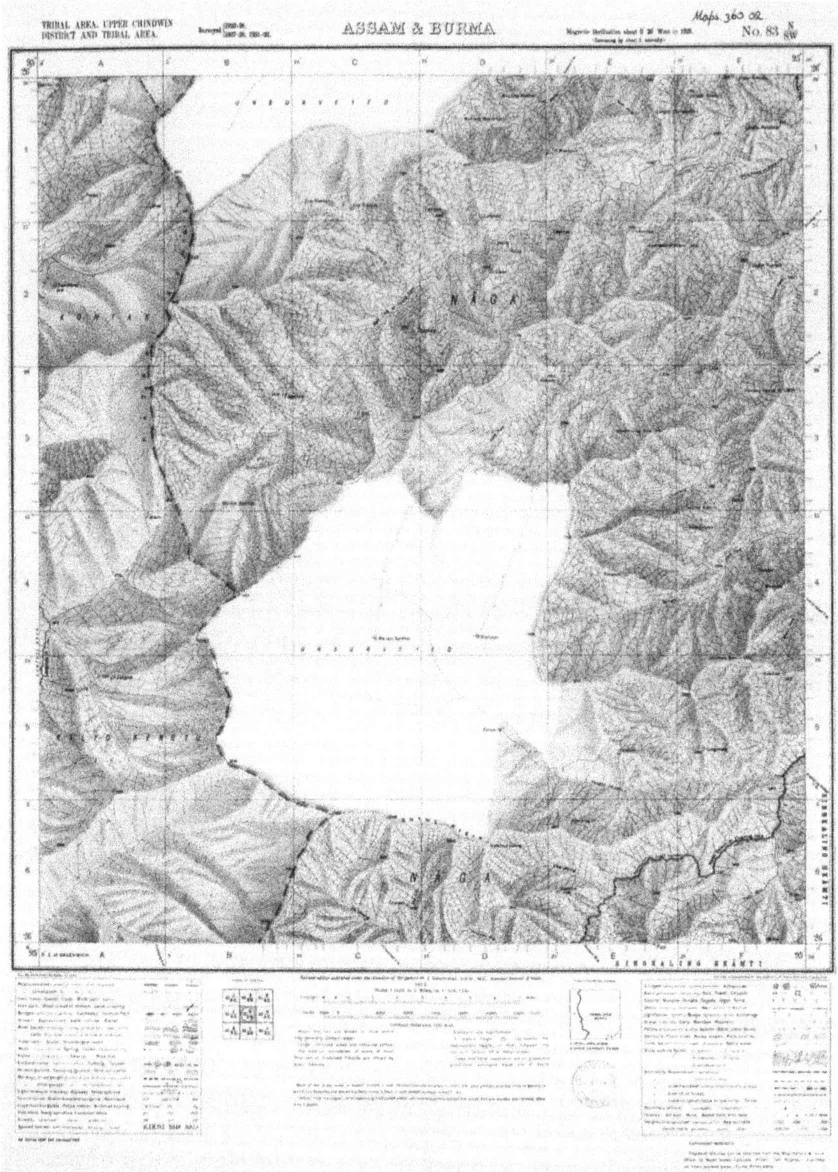

Map 3.1 Topographical maps of Assam and Burma; the blank space is an unadministered and unsurveyed area.
Source: Peter Van Ham and Jamie Saul. *Expedition Nagas: Diaries from the Hills in Northeast India 1921–1937*, 2008, p. 16.

In the paper, Fürer-Haimendorf refers to J. H. Hutton's *Diaries of Two Tours in the Unadministered Area East of Naga Hills* (1929) – tours undertaken in early 1921) – and that Hutton was one of the discussants of the paper. Hutton had been a political officer and deputy commissioner in the Assam Naga Hills (1909–35), where his duties included acting as a local magistrate, settling disputes, and as taxation officer. In 1920, he had been appointed honorary director of ethnography for Assam; he was subsequently appointed professor of social anthropology at Cambridge University. The other discussant, Henry Balfour, was not only president of the Geographical Society but first curator of the Pitt Rivers Museum in Oxford. He had visited Hutton in the Naga Hills and described him as 'the greatest authority on the Naga Hills in general' (Balfour, 1938: 21; Pitt Rivers website). Fürer-Haimendorf's tributes in the paper also include J. P. Mills, the leader of the 1936 Pangsha Expedition, who was an officer in the Naga Hills between 1916 and 1938. Mills, too, published extensively on the Naga people. He later became a reader in anthropology at the London School of Oriental and African Studies, where Fürer-Haimendorf himself would go on to become a professor.

This event at the Royal Geographical Society in 1938 therefore assembled the key 'administrator–anthropologists/ethnographers' of the Naga Hills, who produced extensive ethnographic descriptions of the tribes they visited in the course of their tours through the region and collected a vast number of local material cultural artefacts, many of which made their way to the Pitt Rivers Museum. Fürer-Haimendorf represented the new generation of professional anthropologists. Also, while Hutton and Mills were voracious in their collection of material cultural objects, Fürer-Haimendorf's hunger was for photography. Indeed, a year later, he published *The Naked Nagas* (1939), a book oriented to the popular market rather than being an academic study, and presented as 'the impression of a Western observer exposed for the first time to close contact with an Indian tribal people persisting in an archaic way of life'. This book was highly popularized through a series of articles in the *Illustrated London News* and extensively illustrated with photographs. However, like the preceding generations of ethnographer administrators, Fürer-Haimendorf later assumed a hands-on role as an advisor to the Nizam in the southern Indian state of Hyderabad for Gond resettlement in Adilabad district and later sat on many Government of India committees on tribal welfare (Fürer-Haimendorf, 1939, 1984). He also played a crucial role in the Apatani Valley as a political officer of the North Eastern Frontier Agency (NEFA) administration at the onset of the Japanese invasion of the Naga Hills through Upper Burma in 1944–45 and worked for the colonial administration to draw up plans for the economic development of the Apatanis of the present-day Arunachal Pradesh (Blackburn, 2003; Fürer-Haimendorf, 1947). For Mills and Fürer-Haimendorf, the 1936 expedition allowed unprecedented access to realms no white man had previously visited, raising the veil on terra incognita. However, the expedition and its documentation present a far more complex story than just an adventure into the world of the never-visited Nagas and unmapped land. In many ways the expedition was quite ordinary, much like the routine tours made of the administered areas. It is a good example of colonial ethnographic exercises in the Naga Hills, which not only remained committed to the service of the empire and its military-police-led expansion over non-state spaces, but it also recorded details of indigenous material culture from the style of buildings, to the look of the different peoples to the staple crops, all the while seeking to build a network of friendly relations with the chiefs.

What is remarkable is how the descriptions of ethnographic and diplomatic enterprises coexist, in the same texts, with the depiction of the violence inflicted, notably anti-headhunting raids that are better characterized as acts of colonial counter-insurgency (using the term *counter-insurgency* underlines how colonial pacification tours themselves built a new relationship between the Nagas and the British frontier state administration). Such ethnographies were produced within, and yet were distanced from, the realpolitik of colonial empire-building exercises on the Naga Hills frontier. The Pangsha expedition was extraordinarily well documented. As well as Fürer-Haimendorf's paper, there is, of course, Mills's official account in his reports. In addition, there are Mills's *The Pangsha Letters: An Expedition to Rescue Slaves in the Naga Hills*, letters written by Mills to his daughter, Geraldine Hobson (and published by the Pitt Rivers Museum in 1995). No other colonial documents capture in detail a single anti-headhunting and slave pacification raid, with perhaps the exception of Hutton's 1929 *Diaries* recounting his tours undertaken in early 1921. As a consequence of the Pangsha Expedition, a large portion of the unadministered tracts of the India and Burma no-go areas were brought under the Naga Hills administration (see Maps 3.2 and 3.3).

Approaching Violence and Naga–State Relations

Through a brief sketch of colonial intervention in the mid-1930s in the Naga Hills, I shall now open up discussion on how colonial control over the frontier was legitimized on the grounds that the natives lived in a condition of statelessness, practising headhunting and engaging in trading slaves, which were seen as inhuman and barbaric. This chapter focuses on the practice of narrating the colonial violence, disorder and disturbance that were 'business as usual' in the everyday life of the Naga *jhummias*, and Yimchungers in particular. In doing so, I also highlight the Naga–colonial state relations as they

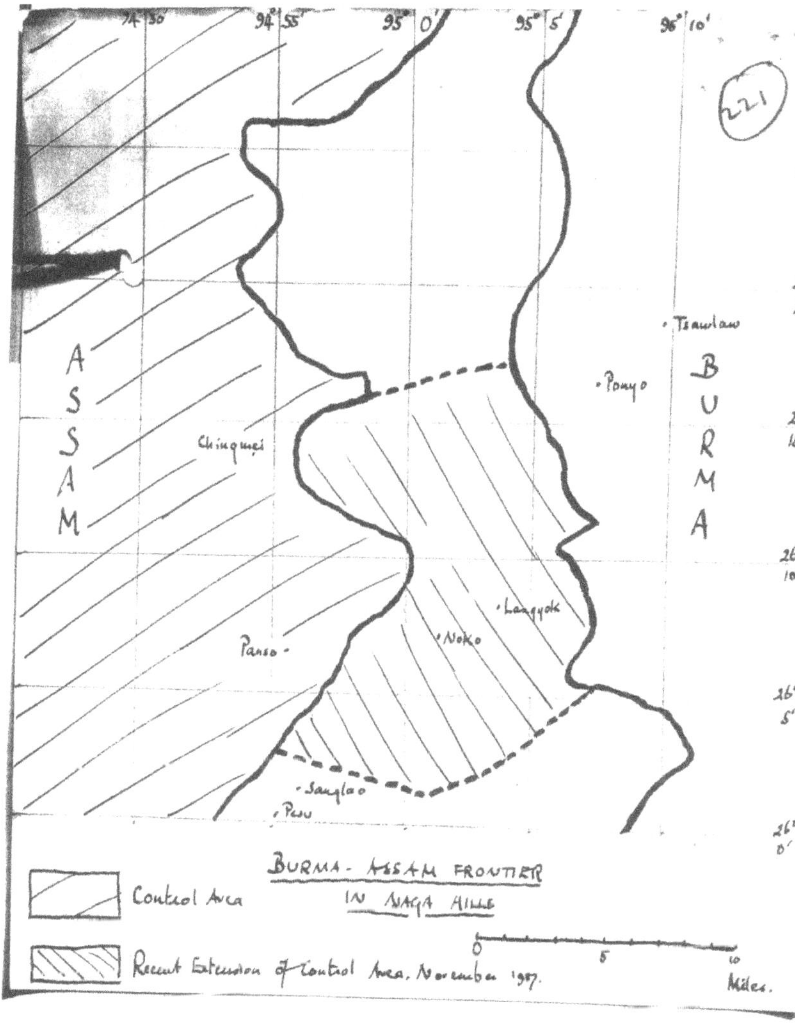

Map 3.2 Assam–Burma frontier map of area brought under Naga Hills in 1937 after the Pangsha anti-slavery campaign.
Source: British Library. IOR. M/3/1457 Frontier excluded areas of Assam and Burma, p. 221.

were built through 'headhunting raids', 'slave capture' and colonial armed tours that conquered the 'savage' races through punitive actions, forced labour and starvation. These strategists of 'civilizing the savages in the Naga Hills' were themselves guilty of acts of violence, which colonial officials termed 'law and administration'. For the colonial administration, the barbarians could only be civilized through brutal strategies of pacification. The contradictions

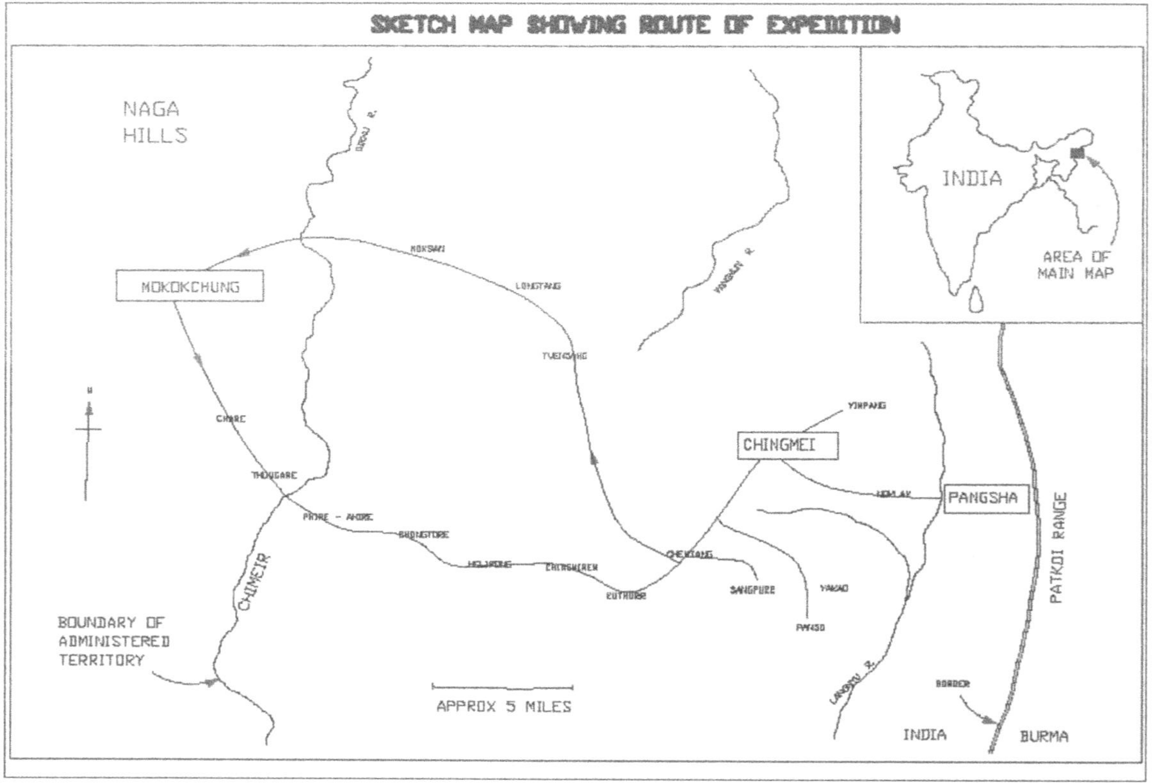

Map 3.3 Route followed by the 1936 expedition.
Source: *The Pangsha Letters*.

are obvious: violence in the name of civilization as opposed to violence as a sign of lack of civilization.

Felix Padel (2009), in his study of the Konds of Orissa, shows how their practising human sacrifice was used as an excuse by the colonial administration to penalize the Konds. To put it differently, Padel questions the very idea of sacrifice as understood by the colonial administration – one that drew a boundary between barbarity and civilization. He further challenges the idea of colonial benevolence, the 'white man's burden', by arguing that the penal system was equally if not more violent than the human sacrifice it sought to curb. Put in the simplest terms, the colonial war of pacification was itself merely another form of headhunting or human sacrifice. In the Naga Hills, native headhunting and human sacrifice were similarly seen by the rulers as 'savage' customs that needed to be vanquished by punitive raids and the imposition of the rule of law. In the administered parts of the Naga Hills, headhunting among British-protected villages led to entire villages being disbanded; anyone found guilty of it was liable to be punished by laws framed by the deputy commissioner's office. In the unadministered part of the Naga Hills, raids were mounted to warn headhunting Naga villages of the consequences. The Naga–state relationship was thus undergirded by violence.

In this chapter, I am using violence as a trope to engage with the colonial ethnologies and administrative writings that were systematically claiming a civilizing mission to 'improve' native backwardness, as documented extensively through presentations of their savage life. I also argue that violence can be used as a trope to understand how state power operates in the margins of the nation state along borderlands. My violent encounter in the field, the hostage-taking incident described in Chapter 2, was linked to this kind of violence that has become so frequent in Nagaland as development grants sponsored by the central government under various rural development projects are claimed by competing stakeholders – from state bureaucrats and politicians to village-level intermediaries and Naga nation workers. They often become a bone of contention between opposition and ruling parties in the state legislature.

Here, I look at the victims' perspective and how they see violence. I try to nuance the whole practice of 'telling' violence by engaging with multiple narratives produced by diverse actors, locating the discourse on violence in the historical context of their writings. Through my narrative and experience with violence during fieldwork, I wish to propose in this chapter that oral narratives (oral history) and written (textual history) both can be deceptive, as the narrators 'agency' mediates in producing the narratives and, hence, both can be mutually beneficial or complementary towards the production of a richer ethnographic description of violence.

During fieldwork I recorded a hymn from the Sanghpur village Yimchunger headmen, who recalled the violent insurrection of 1942, during which they were victims of anti-headhunting expeditions carried out by the British. The context, as narrated by Sanghpur Village Council elders, was that the village was accused of head-taking from a neighbouring village that was part of the administered 'control area'. However, the village elders insisted that this was a false accusation by their feuding neighbours who wanted to take revenge for past unsettled feuds, but were themselves too weak to do so directly. The British punitive tour was their opportunity. Captain Adams, who led the expedition, became the protagonist of the raid. The village elders sang a hymn that captured the sorrow of the raid, in which many Sanghpur men lost their lives and were subsequently driven into the surrounding forest. Such raids were common in the 1920s and 1930s, as rumours and the lack of accurate information often led the British to raid innocent villages. These raids gave British-protected villages the opportunity to avenge themselves on their enemies. Such raids were disastrous as they led to the punishing of guiltless villages.

Inducement to false raids is also recorded from J. H. Hutton's tour, where he wrote that on his march to Tangsa on 6 April 1923,

> the Yacham [Yimchunger] leader came in again to see what we were up to, and also two of the chief men of a village called Shua, or Wakhu, east of the Yangmun, who had heard of our projected tour in that direction. I fancy their purpose was to see if they could induce us to attack a neighbouring village with whom they were unfriendly. (Hutton reprint 1990: 2)

Similar observations can also be found in J. P. Mills's letters to his daughter, Geraldine Hobson, written during the Pangsha anti-slavery expedition.

I recorded the Sanghpur elders' song memorializing their dead:

Sangphpur Kiulung Yankkhi (khun)
Kam Atam. Shaho nu bebah supharu ahihkhuk
Moluh shuh pah sangphu wongtutuntuntearu
Thiyam. Jakam. Taka pahjihoe tush. Mihtsuru lak sahdearu
Yim sangbeak nu tusho thsamu tsonnu
Thusde sangluh shuh pa keochihoo
Abehpuh. Atam-shao lang sangluhnu youri jihd
Sangphpur ahrungru azuru azoru
Ahih khuh laka pahjihdo
Mulung tuk thus lah

Translation
[Sanghpur Village chased out song
Atam [Adams] Shaho carried many soldiers
Armed with guns came to Sanghpur mountains
Houses, granaries burned and people killed
We used tree trunk and grain grinder to protect and fought the enemy with bow and arrow
Commander Atam Shaho injured by our arrow
Our Sanghpur brave man young and old, many people died
Feel sorry and unhappy as we sang.][2]

The hymn, narrated by Sanghpur elders, reflects Yimchunger resistance to state power. The raid was so disastrous for the villagers that it compelled some of them to join their kin in newly established villages, as they had lost so many family members.[3] The punitive raid on Sanghpur was a strong reminder of colonial coercion on Yimchunger Naga villages. Even official colonial accounts describe the aftermath of such a raid quite evocatively; Hutton, for example, describes the scene of a village as it was set to be punished:

> The hostile clan has absolutely cleared out. There was not even a chicken left in the place, and even the houses in many cases were dismantled leaving only the framework. The whole place was crawled with myriads of hungry fleas, the only inhabitants, far too many of whom were brought away with us. Firewood, thatch, even lumps of clay for making pots were all removed, and we came across some of it hidden in the jungle around the village. Even the great wooden drums had been dragged off into hiding somewhere for fear of what we might do to them. (Hutton reprinted 1990: 3–4)

The colonial administrators saw these missions as necessary for bringing peace and order. It is vital to remember that they were focused on raids across the boundary that separated administered and unadministered territory. On one side of the boundary were the British-administered and protected villages that had long given up such practices for their wellbeing and advancement and, by implication, counted as 'civilized' – or, one could say, territorialized. On the other side were the raiders, with their inhuman practices and generic feuding and violence – leading to 'chaos' and 'disorder'. It was at best fleetingly recognized that the colonial 'counter-insurgency' attacks led to the complete decimation of settlements that were burned down and made orphans of those children who survived.

As well as the physical and psychological trauma caused by these colonial raids, they also involved a cultural invasion. This comes up particularly in Harcourt Butler's (then governor of Burma) anti-slavery campaign in Burma's slave district (1925–28). The colonial administration not only used brute force but also symbolic power that measured the Nagas as 'backward' and 'uncivilized'. The use of gramophone, elephants, guns, gunpowder, money and government decree demonstrated modernity and power to suppress headhunting raids, feuds and slave trading among the Nagas. The visual representation of Nagas through photographs since 1920 became a powerful instrument of objectification. Harcourt Butler used these instruments to demonstrate colonial superiority during raids in his campaign against the Kachins, Shan and Naga, which led to the release of hundreds of slaves, the disbanding of human sacrifice and the trade in slaves among the Rangpang Nagas of the unadministered Naga Hills.[4] In order to execute slave-release expeditions, a number of maps were prepared of the unadministered slave district and were coded ('Release and Rescue' maps). The modus operandi of Butler's campaign was to organize a *darbar* (official gathering of chiefs and commoners) at which he declared the will of the government and also paid cash compensation for the release of slaves. This event was brilliantly documented by the metropolitan press in London (1925–28), as the abolition of slavery was urgently pressed for by the Aboriginal and Anti-Slavery Committees in London and by the League of Nations after World War I.[5]

Balfour and Hutton, commenting on Fürer-Haimendorf's lecture delivered at the Royal Geographical Society, were correct in stating that the campaign led by Mills to rescue slaves and punish headhunters, which followed nearly a decade after Butlers campaign in Burmese side of the Naga Hills, was achieved with very little bloodshed (Balfour and Hutton 1938: 283). Nonetheless, it entailed a far more catastrophic symbolic violence. The whole expedition dehistoricizes the practice of slave-trading and head-taking that was central to the local economy but was purely seen as acts of barbarity and exploitation.

In the next section I offer a brief periodization of colonial and post-independence political interference as the colonial administration came to territorialize the Naga Hills as a resource-cum-political frontier sandwiched between Assam and Burma.

Statecraft at the Margins: Colonial Encounter and State Rule in the Naga Hills

Once the Naga Hills became part of the British Indian Empire after the 1860s, violence, anarchy and 'chaos' was a routine of everyday life – as is evident from colonial official reports and also the private correspondence of the military and civil administration deployed to bring order among the 'savages'.[6] Colonial state intervention in the Naga Hills can be recounted in four distinct phases. The first phase began in the mid-nineteenth century as colonial capitalist interests, in opening up land for tea estates in the plains of Assam, began to expand into territories in the Naga foothills, where they came into direct confrontation with the Naga *jhum* cultivators. The Naga in turn raided the settlers, focusing on the tea garden coolies to drive them away from the plantations (Allen, 1905; Mackenzie, 1864; see the full reference in Chapter 1). In the name of protecting tea planters and coolies from Naga raids, the Assam administration in turn retaliated with innumerable punitive expeditions into the Naga Hills. In 1866–67, a portion of the Naga Hills[7] was put under permanent British administration with the creation of the Naga Hills District and, in 1873, while the 'Inner Line' Regulation (explained in Chapter 1) provided for the creation of a boundary between the hills and the plains beyond which non-tribal or plains populations were not allowed to move freely without state permission. Trade was also regulated.[8] It was not long before the discovery of minerals, such as coal and petroleum in Upper Assam, the coming of the Assam–Bengal Railway Company and the creation of reserve forests brought the Naga into confrontation with the colonial Forest Department and the commercial establishment, as well as with the villagers resettled from the plains by the British for forestry work. During this period, colonial military and topographic surveys became the mainstay of colonial information-gathering on the Nagas.

The second phase of violent colonial interference started in the mid-1920s in the unadministered Naga Hill areas on the borderland of India and Burma. This was largely prompted by (or was started under the pretext of) the 1926 Slavery Convention of the League of Nations (which became applicable to the Naga Hills in 1936). The colonial administration of British Burma had already launched a number of expeditions against the unadministered Nagas from the Burmese side to quell the practices of slavery, head-taking and human sacrifice. This resulted in strange compromises, such as the payment of money for slave release.[9] The years following the anti-slavery expeditions were caught up in the British military efforts during World War II. Headhunters were now rewarded for their barbaric habit: military correspondence reveals that the British honoured Naga headhunters with cash rewards for taking Japanese heads during the war, even as they were simultaneously condemning headhunting raids on neighbouring Naga villages (see Image 3.1).[10] The Nagas fought well along with British and American troops and launched a guerrilla resistance against the advancing Japanese forces. In 1947, after the war, these areas were transferred to the Independent Indian and Burmese administration.

Image 3.1 Pictures depicting a Japanese raid in 1944.
Source: *Kohima 1944, The Battle that Saved India*, p. 64.

During World War II, the Naga Hills became an area of unprecedented national and international activity. In 1942, American and British forces were being deployed with the local Naga *dobashis* to contain the Japanese advance through Upper Burma to Kohima. Along with a large number of troops, the construction of infrastructure (road, river and air) involved Naga and non-Naga forced labour amounting to 250,000 workers (Franke 2009: 61). Nonetheless, the military reports and fortnightly administrative reports of that period narrate constant confrontations with British soldiers mitigating feuds, headhunting and human sacrifice. For example, in one of the frontier reports, it is stated that 'the outbreak of head-taking in the southern Sangtham country is reported. The troops which are in the area till fairly recently have left and the Nagas find it difficult to understand why they cannot attack their enemies just as we attacked the

Japanese.'[11] In March 1944, the Japanese launched their offensive from Burma and, fighting their way westward into the Naga Hills had, by the end of the month 'encircled 155,000 British and Indian troops in the Imphal area'. In early April, they reached Kohima and besieged the garrison' (Franke 2009: 62). The battle of Kohima lasted nearly three months until the Japanese began to withdraw on 31 May. Franke cites the view that the four-month Japanese occupation 'effected the abolition of the arbitrary division of the Naga Hills by the British between administered and unadministered areas' (Franke 2009: 62).

The third phase of violent colonization began in the years after 1947, which saw massive militarization of the frontiers by Burma and India. A new era of counter-insurgency operations began in the early 1950s and continues to this day. During this period, the Village Defence Guards (VDG) were raised by Major Ralengnan Kathing in Tuensang District to protect Naga villages from the insurgents. The Village Defence Guards have, since their inception, played a very important role in the counter-insurgency operations. They acted as mediators, go-betweens and as participants in conflicts that unfolded. To cite one example, during the days of the Naga National Movement, the Village Guards played a crucial role acting as a safety valve between the village and the resisting underground forces. They at times acted as a conduit to violence and as sentinels of the armed forces as a village-protection force. To quote from one official source of the Government of India, 'The primary role allotted to the VDG [Village Defence Guard] was the immediate security and protection of their own village against violence to life and property. Secondary roles were: limited protective routine duties such as road opening, passing of information, acting as guides and interpreters and provision of escort to civil Administrative Officers.'[12] During my fieldwork at Tuensang district headquarters I recorded from the Village Guard[13] staff stationed in Tuensang town that they are these days assigned additional duties during elections to protect ballot boxes from being stolen by opposition party people and are used as escort attachments to safeguard local politicians and Members of the Legislative Assembly (MLAs) candidates during their election campaigns.

As late as the 1980s, over a conflict of land in Tabu, claimed by both the Chang and the Konyak Nagas, the Village Guards from the respective communities were used as go-betweens to contain feuds by the administration. In my own fieldwork, I recorded from my interlocutor the important role played by the Village Guards in countering the Yimchunger–Tikhir conflict. The Village Guard played a very crucial role in protecting the interests of the state, sometimes far greater than the Assam Rifles (AR), the Indian Reserve Police (IRP) and the Nagaland Armed Police (NAP) engaged in the counter-insurgency operation in present-day eastern Nagaland. The period saw state violence and counter-violence from Naga nation workers who were demanding sovereignty.

The government, on the other hand, handled the situation as a law-and-order problem that nonetheless had deep undertones of 'ethno-nationalism' (Lotha 2009). In a 1956 newspaper article that first appeared in the *Calcutta Statement*, Prem Bhati writes:

> Since war, not peace, makes most of the world's important news, it is not surprising that we usually think of the tribal people along the North East Frontier only during periods of emergency, such as the one that now threatens the Naga Hills District [...] This may be a regrettably popular attitude, especially as the tribal areas in this part of the country are going through a process of unprecedented development, but it will be folly to ignore the recent incidents. These represent not only a law and order problem but are also symptomatic of certain political conditions and a consciousness to be free. Nevertheless, seen in perspective, their importance is limited.[14]

Since the mid-1990s, there has been a renewed interest in nationalizing the frontier space through decentralized governance and with massive central and external donor assistance. This initiative of decentralized governance and aid economy has established a new equation in villages resisting state power. It has resulted in factional clashes and kidnapping and intensifying revenge killings and extorting cash from people who, by managing such schemes, had become their main beneficiaries and hoarders of public funds and grants. The new wave of violence has also been caught up in electoral politics, with massive vote buying.

What connects these four phases (colonial, World War II, post-independence North East Frontier Area administration and the present ceasefire) are the visual and textual representations, often expressed by post-independence administrators, of the Naga as the 'violent other' through images of the Naga frontier as an 'out-of-the-way' place (Tsing 1993).[15] The visual images contribute to the oral and textual discourse through which colonial and post-colonial administrations have justified their actions in the name of quelling violence and bringing 'law and order' by pacifying headhunting forays through retaliation raids. The finest documented evidence that lives in public memory as a slave-rescue expedition, claiming colonial benevolence is provided by Pangsha village. While local people were accused of head-taking, human sacrifice (practised by the Rangpang Nagas according to Burma colonial sources) and slave trade, they saw these practices as central to their social organization and crop production.

Structural-functionalist explanations by Hutton (1923) of head-taking rituals describe the practice of head-taking forays as the basic need for fertility and more importantly to gain manhood and status in Naga society. However, according to the colonial administrators', this practice needed to be dismantled for the progress of mankind and humanity. For my Naga interlocutor, who narrated the Sanghpur raid of 1942 through the British campaign song, the government forces were villains and the perpetrators of violence against innocent life. The administered Nagas who conspired against them and helped take heads were, as Anna Tsing would call them, 'government headhunters' (Tsing 1992). The narratives of colonial state violence against the Nagas were, I later found out, even documented in these areas.

On 14 August 2009, I was invited to join the Naga Independence Day celebration in Chassir (Tsasher in Yimchunger) village by the Naga Nation Workers from my study village. In the booklet that was circulated during the day-long function, I could gather references to Sanghpur, where the first Naga resistance to colonial invasion of their territories was made. The Naga Nation Workers described the British raid as 'invasion' and Naga opposition as the 'first war of resistance'. They were keen to establish that it was the first war of Naga independence. The booklet was issued in the name of Y. Lakham, who is regarded as the Naga warlord and was a comrade to Angami Zapu Phizo during the heyday of the Naga National Movement. From the narrator's perspective, the people who came to settle feuds and pacify headhunting forays were the perpetrators of violence. I felt that some of these people even exaggerated the Sanghpur raid, claiming that their neighbouring administered Nagas, who connived with the British and directed the raid against them, boiled the flesh of victims and took the heads away as trophies to the administered area. The village elders were terrified and full of sorrow as they narrated the violent event and sang the song, which brought tears to their eyes.

The narrators had a deep sense of unity and brotherhood in their hearts. Their emotions were high, but they also acknowledged that these days, such spirits have become corrupted, with underground factional feuds and ongoing peace negotiations with the government. Greed for money, power and wealth had dissuaded people from the path of nation-building. On the next day there was a grand event in Shamator town headquarters: the Indian Independence Day celebrations. As I attended the state-sponsored function, I realized how both resistance and pacification are blurred by 'state rule'. The local MLA, who attended the subdivision headquarters and functioned as state guest of honour, read out his party's achievements and announced a number of packages and programmes that the government were implementing for the people of the district. A massive paramilitary presence surrounded the Shamator town playground, guarding the parade and cultural festivities. The programme was also attended by local government dignitaries, Baptist pastors from neighbouring villages and Shamator town Catholic priests. My village pastor had worked hard over the year to present his troop of young children, who were raised under the National Social Service scheme and Bharat Scouts and Guides. The parade ground was also ringed by a number of stalls displaying to the public the progress of various community developments and state-sponsored schemes. The state Agricultural Department had set up an interesting stall that presented the genotypes of long beans that had been made popular by the study village (Leangkonger) in the district. The Indian Independence Day celebration was a confirmation that development was the last battleground that communities now face as a violent encounter to protect their livelihoods or rather to reorient them to market forces that were fast-changing. The coming of cash through central- and state-sponsored schemes has created new constituencies of power that create conditions of friction at the local level and often result in violent struggles as discussed in the context of my fieldwork in Chapter 2.

The out-of-the-way places have increasingly come to be identified as places that are in an active process of engagement with state power and its politics of nationalizing frontier space (Baruah 2007).

Nationalizing Frontier Space

This section looks at how violence emanates from the friction between expanding state power and the state-making practices carried out at the frontier, focusing on the tour diary of a postcolonial military official, Man Bahadur Rai, who served as the Frontier Administrative Officer in Tuensang Frontier Division during the 1950s, as assistant political officer, and in 1957 as assistant commissioner of the Naga Hills Tuensang District.

By 1947, when the British transferred power to the Indian government, the territorial boundary between India and Burma still remained undecided – the last unsettled frontier of the empire. In 1937, the administration of Burma had been separated from India, and this also marked the beginning of the division of the Naga Hill areas between India and Burma. There were frequent debates about Naga Burma and Assam Burma in 1930s colonial correspondence. In these late colonial discourses, the Nagas evolved into a people who were intended to be part of a nation state, incorporated through tour expeditions and frontier map-making by colonial officials from both sides of the frontier. By 1940, Colonial Roger Mitchell, in his tour report on the Naga Hills, had declared that all unadministered Naga areas in Burma had been brought under political control. However, these were largely 'paper' claims as the ground realities made it increasingly impossible to administer these areas: colonial administration could penetrate but not stay for long. The Naga National Movement, like other indigenous national movements, had already begun articulating its aspirations during this period. On the eve of

India's independence, all the unadministered Naga areas were brought under the territorial control of NEFA administration. Intermittent guerrilla warfare had begun between the military and the Naga Nation Workers, in those days called the Naga National Council (NNC).

Anthropologists remained completely silent on these political developments, and a distance was maintained in the ethnographic writing. In the 1960s, Fürer-Haimendorf, who managed to get limited access to the Naga Hills, clearly states in a paper in *The Geographical Journal* that unlike during his first fieldwork in 1937, he would keep himself away from the political discourses that were shaping the relationship between the state and their subjects in the Naga Hills. Similarly, Milada Ganguli, author of *A Pilgrimage to the Nagas*, entered the Naga Hills 12 times in those three decades but was completely silent on political issues. Anthropologists were now collaborating in the making of the postcolonial nation state, advising the national and state governments on the welfare of tribal areas. This uncritical engagement of colonial ethnographic writing became a handmaiden of the empire's administration and continued into the post-independence period. But now the relations between the state and the semi-nomadic Naga swidden-*jhum* cultivators were shaped both by more intense political and administrative control and by the struggle for self-determination.

On the eve of India's Independence in 1947 (once dominion status was granted), the frontiers became the most important sites of contestation. During World War II, fortnightly reports were sent to the central government in Delhi and presented a new territorial (and violent) representations of the Nagas. A report in 1947 reads:

> The Government of India has complained that Nagas from Burma, in particular from Ponnyo, Tsawlaw and law Naukkon, have raided territory in Assam and have taken, in an unspecified period, over 700 heads. Villages raided in Assam were Ukha (400 heads), Pochu (110 heads) and Yonghong (200 heads); in the case of Yonghong, the Tsawlaw Nagas from Burma were called in by Sholo (a village in Assam) to help their war against Yonghong and the result was the loss of 200 Yonhong heads. As a result of the evacuation of British and Chinese troops through this area in 1942 the Nagas of Ponnyo, Tsawlaw and Law Naukkon have acquired modern weapons which have enabled them to undertake these very extensive raids. The government of Assam is desirous of sending a punitive column in Burma territory to punish these villages.[16]

In the preceding years, the practice of headhunting had declined as the state administration and Baptist missionaries established their presence in the hills, but the relation between the state and the semi-nomadic Naga swidden cultivators was now shaped by more intense political and administrative control and struggle for self-determination. However, with the increased security of the area, feuding headhunters now fell prey to government soldiers carrying out orders and quashing dissent in the newly launched counter-insurgency programmes. The Nagas were seen as part of the nation states, and headhunting and slave-taking were now cast as a national-security concern. The statelessness of the Naga world was quickly fading away as they were incorporated within the nation states of India and Burma.

Rai's appointment as assistant commissioner and his tour took place in the context of emergency measures and militarization. The Assam Maintenance of Public Order (Autonomous Districts) Act 1953 and the Assam Disturbed Areas Act 1955 were in force. Shortly afterward, the Armed Forces Special Powers Act 1958, and the Nagaland Security Regulation 1962 would add further legal sanctions for army operations. Tuensang Division had become an active centre of insurgency by 1955, with paramilitary units and the army called up in 1956 as the region was declared a 'disturbed area'. Rai himself was an officer of the 8th Battalion, Assam Rifles, from 1957 to 1962.

Yet, at the same time, Rai's 1957 tour report seems remarkably like a normal tour diary; like Mills's report of the 1936 Pangsha expedition twenty years earlier, violence is combined with regular observations. Rai's tour report presents fascinating insights into postcolonial state incorporation that redefined slave takers and slave hunters as victims. His tour notes, like those of his colonial predecessors, identify who were friends and foes among the Nagas, thus continuing the strategy of colonial territorial incorporation through producing bifurcated classes of people within Naga communities. The trusted natives and the wild savage, the more exotic, and 'barbaric' Nagas, 'untouched by civilization', occupied the untamed frontier on the Patkoi Range along the Burma border, while the more 'civilized' Nagas lived within the administered tracts of the Naga Hills. The colonial need to bring order had now become a question of national security. Rai, a Gurkha official who had served in the British Army, was made assistant commissioner of the Naga Hill Tuensang Area in 1957, which had been declared a Union Territory that year. His tour diaries, also like those of his predecessors, give a valuable account of how native practices of farming, livelihood and village-level administration were seen by the state in this frontier area. The chase was now not for people who took heads, but to disarm the resisting Nagas through political patronage and military control. These tours also reflect the mentality of postcolonial state officials as they produced their reports for the government (NEFA administration). Rai's tour report was a regular annual account of frontier administration to settle disputes in the unadministered areas and to collect information on hostile villages.

Man Bahadur Rai's annual tours were like those of his colonial predecessors, part of regular frontier administration, and collecting information about hostile villages – although now focused on villages that were sheltering the Naga nationalists and rebel leaders. Rai's colonial predecessors made innumerable tours in these unadministered territories to suppress headhunting, raiding and slave-taking. In the postcolonial period the practice of headhunting continued but was declining; tours and official visits were now aimed at extending the general administration into these inaccessible frontiers. The tours were now accompanied by new sets of interventionists who accompanied the assistant commissioners – as well as the *dobashis*, without whom these tours were near nonstarters. Also, there was a new development: village-level grassroots agricultural planners, people who worked as agricultural demonstrators and workers started accompanying the touring parties. While the emphasis, in the colonial period, was trying to establish political and administrative control and pacification of the unadministered villages, the post-independence administration had established permanent control over the frontier and thus turned its attention to improving native life by introducing new farming techniques. Promoting settled agriculture became the immediate priority. Even village settlements were becoming more permanent as hostility and headhunting forays between feuding villages declined under government and missionary presence.

In addition, there were the newly formed Village Defence Guards, raised in 1957 as a measure to ensure village-level security in the area (Image 3.2). The colonial administrators were very ignorant of the unadminstered Naga Hills, as they only saw fields in the forest landscape that immediately mirrored natives' primitive vagabond habits. These tours were also critical to collecting information about raids, hostages and espionage against the state. The colonial officials brought violence by raiding innocent villages and taming local chiefs by containing raids. The postcolonial government raiding parties used national security and territorial integrity as rhetoric to pacify resistance.

Like the tours conducted by colonial officials, these also involved adjudicating, on the village level, disputes over land and territory. What made these dispute settlements more interesting is that, when they involved transborder tribute giving, they were now mediated by political administrators on both sides of the border (India and Burma). Postcolonial mediation made these blank, 'out-of-the-way' places legible as national frontiers, but at the same time, cast the inhabitants both as people belonging to a wider political context and as marginal, minority and borderland people.

Image 3.2 Major R. Kathing and his wife with early village guards.
Source: Village Guard Nagaland Souvenir, 2007, p. 23.

This section continues to explore the trope of head-taking. For postcolonial administrative officials, as for their colonial forebears, headhunting was vividly condemned in their narratives as evil, characterized as lawless, savage violence. While raids and burning villages in search of hostiles and rebel parties were seen as regular practices of administration to tame the wildness and produce law-abiding citizens, their own practices were not recognized as violence but were perpetrated in the name of civilizing the frontier space. The geopolitical location of these frontier spaces presented a unique challenge to state power. The post-independence Indian government had defined its territorial control by seeking to demarcate a definite political boundary with the Burmese administration. In this context, rebellion, organized violence, and arms traffic across the border were seen as threats to national integration. The local space had now become part of the nation, and frontiers were fast becoming borderlands with well-defined boundaries between states that demanded vigilance and control over frontier people's lives. Even the way of farming, swidden, with its shifting sites, was seen as an evasion of state control (Barua 2007; Scott 2009). Now I narrate in brief the 1957 tour.

The 1957 Tour

On 12 June 1957, Man Bahadur Rai records in his tour report that he had received information that the assistant resident at Lahe in the Burma Naga

Hills had informed his office that he would be arriving at Pangsha (the same Pangsha of Mills's 1936 anti-slavery expedition). The next day he was expected to arrive at Noklak, the administrative outpost, for a meeting at which fines and compensation paid by Ponnyo village in Burma to Nokhu village in Naga Hills Tuensang Area (NHTA) would be handed over by the representative of the Burmese administration. The incident clearly demonstrates how local relations between raiding villages were now being regulated by nation states, asserting authority and control over native villages that were now located in the borderland. The tours had the objective of pacification as well as giving tribute to villages that cooperated with the administration.

The previous day, when the Assam Rifles party had reached Pangsha village, the RMO settled a number of disputes that had been pending in the village. The commandant had given two mounds of salt and 30 seers of rice as a political present to the village headman. However, villages that were supporting 'hostiles' (NNC sympathizers) were systematically punished by being burned to the ground. For the native Naga villagers of the NHTA, such a two-pronged approach of appeasement and intimidation was their regular experience with state authority, through which Naga citizenship and allegiance were constituted in the post-independence state.

On 1 August 1957, Rai reached Phombak, a suspect village, with his troop of Assam Rifles and 30 Village Scouts. The objective was to find hostile camps in the vicinity of Phombak. As he reports:

> They [the Village Scouts] returned back to the camp at about 1800 hrs. They said the camp was already destroyed by one of the operating parties, most probably Longling. No one was found around the camp. It was very strongly fortified with Bamboo spikes and stockades constructed out of huge stones with loop-holes and slit-trenches. When our party reached the spot the camp was still burning which indicated that it was destroyed only two to three days back.[17]

Rai's party arrested one man, Thangboh of Phombak. According to him, the hostiles had deserted the camp about a week before it was raided. Phombak was one of the villages which supplied foodstuff regularly to the hostiles in the area. For this reason, Phombak was burned down by Rai's party on 30 July.

After being fired upon in Nakshu village, Rai's party launched a fresh search for the hostiles, who had become invisible fugitives in the dense tropical jungle, and made an elaborate patrol hunt with the Assam Rifles and the Village Scouts. Rai's account continues:

> About a mile from Nakshu, towards due north, we discovered a small jungle track which bifurcates from the main track and debouches into

> deep jungles. After about one and a half hours search in the area at least some of our village scouts located the camp and informed us. Smoke was coming out of the hut and a few people were seen moving about. We were sure that these were the same hostiles who had open fired on our party that very morning. Our party opened fire at them, as a consequence of which three were killed. There were three big huts which could roughly accommodate about 30 to 40 persons. The camp must have been used very frequently by the hostiles whenever they visited Nakshu and Longomg village.[18]

On 27 July Rai left for Noklak from Tuensang with 50 Khemyungen (Kalyo-Kengyo) Village Scouts and three *dobashis*. Their movements were kept secret and, on 29 July, they arrived at Nakshu where they found the camp already prepared for rest. Rai notes that 'the whole village was burned down except ten houses, which were spared for the accommodation of the party'. The village, after being under siege by the Assam Rifles and the Village Scouts, was demolished on the same day at 03:00 hours before the tour party could reach it. The government forces had discovered one light machine gun and three rifles. The village was empty except for two families, who were found dead.

On 10 March, as Major Rai was moving from Taknyu to Pangsha via Samnyu, that is, from Konyak villages through Khemyungen and Yimchunger territory, he met a suboverseer, Sri Yanaosou Lotha, who was living in a headman's house. He had been sent there to find out the alignment for the Noklak–Pangsha road project. The alignment had reached the other bank of the river Lang. Through mobilizing troops and building drivable roads, the state was extending its presence more formidably in these frontiers. On their way to Taknyu (Yimchunger area), Rai describes the site of Impang village:

> Nowadays this village is completely deserted. One can only find the remnants of the once thickly-populated village, the only remnant which still speaks of the ghostly scenes of headhunting by Pangsha that happened only last year. I think the village has disappeared for good and no other village will ever stand on the unfortunate hilltop of this Impang village.[19]

After arriving at the Yimchunger village of Taknyu, Rai's party camped two furlongs away, both for security reasons and also as they were not familiar with the village. That evening villagers helped the party build their makeshift tents with palm leaves, mats and bamboo. In the evening, Rai explained to the village headmen and others about hostile activities. He also warned them of very serious punishment if they themselves helped the hostiles, directly or indirectly. The next day it was time for the marching party to leave and make a surprise visit to Waosoi, a Sangtham village. On 9 March 1956, they reached there early in the morning, without any prior warning. A Peso interpreter had accompanied the party, but Wantu Toshimong, who resided in the village, never knew when they had left for Waosoi. People were frightened of the government troops, and there was general pandemonium. In fact, the officer in command (OC) had divided the troops into three parties to enter village from different directions, with a fourth party, consisting of a few Assam Rifles, accompanying the main body. The villagers had no chance to escape. The interpreter of this village, Chasimong (a *dobashi*), was also present. Rai called him. Nervous at Rai's sudden visit, he came running to see him. Rai asked him to tell his people not to run away and that, if anyone tried to escape, they would immediately be shot dead. Then he called the village headmen and went around the village. Seeing that Rai's party would not take any action against them, slowly the villagers recovered their confidence. In the course of his short stay, Rai explained, as usual, to the villagers the adverse effects on the village of the false propaganda carried out by the hostiles. Since the village was under Noklah (a Sangtham-dominated area) jurisdiction, Rai warned them of severe punishment if any adverse reports were received against the village.[20]

On 7 March 1956, Rai proceeded to Ongongsang village via Niniyan village and then to Chingmei. On the way, they encountered three isolated settlements inhabited by three *jhumming* families named Hangso, Thongeho and Mongba. Rai's *dobashi*, Toshimong, informed him that previous visits by political officers had not had any effect on these settlements. The *dobashis* and the village headman, Wasio, had raided this settlement and burned its houses in the past, but its people had rebuilt them in higher positions. Rai further writes that, upon enquiring, the village headmen and *dobashis* explained that they had purchased the land from Tobu village and had no wish to return to it. They also informed Rai about NNC movements in the area. For Rai and his party, this was a dangerously out-of-the-way place, where people from Chingmsi, Waossoi, Niniyam, Taknyu, Tobu and Hokechang could easily gather for a few days to conspire together and plot their revolutionary designs against the state.

Rai evidently relied entirely on the information given to him by the local village headmen and *dobashis*. Their description of NNC movements, and the hospitality shown towards these men, was enough proof for Rai to arrest them for supporting the hostiles. Rai could not arrest the village headmen and the others without substantial proof. He even needed the help of the headmen to nab the culprits. In other words, it was unwise to prosecute government supporters and informants. Simultaneously, a patrol party was sent around the area to hunt out enemy camps, but they found nothing. Moving on towards Ongpansang, at around 4:45 a.m., the Assam Rifles jawans launched a light parachute, signalling their arrival at Ongpansang. The government 'headhunters' (Assam Rifles) were now let loose for their regular search and mopping-up operation of suspects in the area. The operation did not result in the successful capture of any suspects, and the tour party marched on towards the base camp, ending Rai's journey.

The establishment of new villages was not unusual in the Naga settlement pattern. Indeed, it was the regular mode of migration and movement. A pioneer would move away with a group to establish a settlement, formally marking it as a new village by gifting *mithuns* (*Bos gaurus frontalis*, semi-domesticated cattle) and *daos* (machetes). In my study area, villages were established right up to the late 1940s for reasons, as elderly villagers narrated, of feuds, raids, growing numbers, or nearness to new agricultural fields. In earlier headhunting days, other villages would then raid new settlements, as they saw them as competing for resources. In the past, the study village had been raided several times by faraway villages, whose names the villagers declined to give. As the colonial and later, postcolonial, administrations extended their control over these territories, government officials feared such newly established nucleated settlements, which not only evaded state surveillance but became the remote sites for rebels to design their insurgency against the state.

The political intermediaries along with the allied chiefs had their own motives. As described above, the villagers who were arrested by Rai had previously had their own villages raided and burned down by local *dobashis* and the headman of a neighbouring village. Thus, the post-independence world reproduced the problems of the colonial era. State administrative violence often reinforced (wittingly or unwittingly) family feuds and inter-village conflict. State control over borderlands was thus mediated via the village intermediaries, who played a very important role in negotiating claims but in many cases perpetuated violence – settling their personal feuds by directing government forces against innocent villages.

What occurred was the product of the interaction between regional/local alliances between villages that were, in turn, mediated by the political go-betweens on which the administration depended for successful pacification. However, the point is not simply that colonial and postcolonial officials were duped into becoming 'players' in local internal feuds. It is equally about the lure that external state and governmental objectives and actions offered to political intermediaries (*dobashis* and headmen). It is critical to understand this interaction. It led to much co-production of violence and the state co-option of local divisions – and it would go on to have massive negative effects on Naga development, as described in the later chapters.

Thus, even the location of settlements within villages was challenged by the touring parties, and often they ordered the village headmen not to relocate their wards (*khels*) without government orders. The government soldiers would punish the hostiles (people who had joined the NNC) by warning villagers not to allow hostiles to build houses or settle in the villages. In every village, a household count was done as part of the enumeration to know the size of the settlement. Rai describes his Pansu visit on 15 February 1956 in his tour diary. During the visit he warned villagers, particularly in the Lamtheo *khel*, not to allow such persons who were still in the jungle or otherwise, directly or indirectly, to help the hostiles. He further writes:

> I explained to them their foolishness to believe in the hostiles. It seemed to me that the hostiles had very strongly injected in their mind of American help in getting independence. It was very difficult to explain to them how far America was from them. I explained to them that our government's intention towards them was a very generous one. I told them the hostiles were only a constant source of harassment and trouble for them unless they tried to disassociate them.[21]

In 1947 all foreign (largely American Baptist) missionaries had been expelled by the post-independence government and American involvement was constantly romanticized in policy circles as a potential threat to the nation's integrity. The Chinese were also frequently seen as involved in promoting Naga separatism by supplying arms. Such talk was also circulating among local officials, as they perceived that local people were being influenced by stories of external help in their struggle for self-determination.

But there were also reasons for Rai to be happy. He found a friendly village headman in Pansu and another in Kengnu who had come with their families to live in this village, while another headman had a son in the Assam Rifles. These were very positive signs for the government, which also constantly sought to reinforce loyalty among supporting villages by extending all possible protection from hostile forces – and by providing guns and ammunition to villagers who had engaged in hostilities against the NNC. As Rai writes on 21 February 1957: 'We went around the village [Pansu] with political officers. We distributed 150 rounds of 303s [rifle bullets] to the villagers who had captured hostile arms.' On 18 February, Rai had visited the village and distributed 500 gunny bags to villagers whose houses had been burned by the hostile army and four bags of rice to the village headmen whose houses had been burned down. In such ways, the *jhum* cultivators were slowly made into victims of state design aimed at maintaining 'territorial integrity' on the nation's frontier. By dividing villages into friends and foes of the state, just as in the colonial period, the swidden farmers were made natural enemies of one another (as government supporters 'regrouped villages' by the administration and as branded rebel villages that hosted the underground soldiers and resisted government regrouping plans). The villagers were no longer solely at war with each other, but were working for or against territorial state control.

In the preceding years, the Village Guards, raised to safeguard Naga villages from hostile forces, had been used to aid a peaceful election; various political parties used guards to support their political interests. During my fieldwork, many villagers commented that the guards were now even engaged during ballot-counting, controlling booth-rigging and dealing with miscreants who disrupted the election process. The guards were also powerful in lobbying villagers to cast votes for particular political parties and were implicated in petty crimes and building tensions between villages. Man Bahadur Rai narrated on 19 June 1957 one such incident in which a burglary was carried out by a Sanghpur Village Guard, who was fined Rs 50. Other guards also took an active part in killings in Yakkao village, for which they were punished. In the study village, the guards had stirred animosity with neighbouring villages about the forced clearance of forest areas. Thus, we see again that state strategies of controlling subjects were leading to inter-village rivalry and petty disputes triggered by Village Guards' misuse of power. The guards played a vital role not only in controlling violence but often their high-handedness also triggered disputes and animosity between villages (Image 3.3).

'Government Headhunters'

Anna Tsing (1996) talks at length about 'government headhunters' and how their image shaped politics and projects for development in Indonesia. Under Suharto's regime, headhunting had turned into a way to supply heads during the construction of dams and bridges for construction rituals. In my childhood I had heard these stories from my parents, about government officials ordering heads from the hills just before the construction of big bridges and dams. My mother used to tell us stories that when the mighty Brahmaputra was spanned in the 1950s, several heads were taken from the surrounding hills and human sacrifices were offered for fertility and success. In the field, I never heard these stories.

Tsing's account is not so simple. She is describing how villagers constructed state forces as headhunters in the form of mirroring. As Rodgers summarizes Tsing, 'Kalimantan societies [...] and the nation states that now encroach upon them tend to phrase their relationships to each other in terms of predation, vulnerability, attack and death – in terms of headhunting' (2001: 42–43). Although showing the persistence of headhunting in the Naga Hills in

Image 3.3 Arms captured from the first China-returned Naga rebels.
Source: Nagaland Village Guard Souvenir.

the twentieth century, these commentaries need to be contextualized in the light of contemporary ethnic and territorial conflict between the Naga tribes. These headhunting events at Tobu have a quite different context from earlier periods. They were instigated by territorial and ethnic disputes between tribes who claimed the same territory. The land disputes between the Chang and Konyak Nagas had intensified during the 1980s and had unleashed a major headhunting event that desolated Konyak and Chang village and also engulfed the Sangthams.

My own enquiries about the Tobu conflict remained messy throughout, as it was too complicated to get a clear perspective on the claims and counterclaims over the land dispute between Chang and Konyaks. Evidently there was some fabrication, and matters were also complicated by state policies and practices of arbitrary boundary demarcation that became a norm of colonial administrative control pursued by the postcolonial government.

However, one fact becomes very clear from reading government publications: the state patronized headhunting in the guise of maintaining peace and to establish control over the community. Certainly, during World War II, Naga soldiers were rewarded for taking Japanese heads on the Assam–Burma frontier.[22] This military tactic was also followed in Malaysia, where the British frontier administration was losing control over highland people. Major Tom Harrison, a British anthropologist, was deployed to counter the Japanese invasion, and he encouraged and trained Kalibet highlanders in Borneo to take heads, as described, for example, in the film *The Naked Anthropologist* narrated by David Attenborough. Head-taking was represented as an act of bravery and was used to banish resistance. It was the most violent practice used to victimize the natives. However, this practice continued after World War II. In the 1950s, the Village Guards were involved in building peace through rewarding headhunter guards who were none other than government agents in the conflict (Naga villagers recruited as Village Guards).

In the 1970s, when Tuensang district was divided in two, Tuensang and Mon, conflict arose between the Changs and Konyak Nagas. Both tribes had been partners with the colonial administration much earlier than their Kalyo-Kengyo (Khiamniungan) neighbours. The Village Guard were let loose to capture heads in the Tobu conflict in the name of peace building and successful cessation of conflict that lay in state making and territorialization of tribal identities that came in conflict with the natives' living space. The Tobu conflict brought the state to centre stage of local land disputes and intertribal claims over territory.

In most parts of eastern Nagaland, territorial rights have been settled based on the influence by the colonial administration – and hence by those who influenced the administrators. That explains why, for example, many of the Yimchunger villages are today Changs and vice versa. This dominance of a particular clan over their neighbours was mediated by the influence of political go-betweens who were instrumental in delineating boundaries. In Tobu, the land was claimed by both communities, and the new district boundary became a contest for claims and counterclaims. The Village Guards were the government force that took an active part in the conflict to normalize the situation. From the guards' own publication, their heroic seizure was linked to many stories around Tobu and the pacification raids in which many heads were taken and after which the guards were rewarded by state officials for their bravery and labelled as patriotic 'defenders of national territory and integrity'. In this context, government headhunters were seen as another strategy of the government's administrative apparatus.

The Inseparability of Violence and 'Normal' Government

Colonial administrators outlawed headhunting and human sacrifice to extend their administration's control over the unadministered Nagas through punitive raids and laws that were considered humanitarian, moralizing and necessary to civilize the natives (see particularly Roque 2010). The postcolonial government embarked on similar developmental missions by spreading the message of welfare through various development schemes that were used as instruments for winning over the natives and isolating the population by 'protecting' them from what hill administrators claimed was the false propaganda of Nagas seeking self-determination. As one NEFA file, written on the regrouping of Naga villages, illustrates:

> Frantic propaganda is being arranged by some of the Ao hostiles to persuade the Konyaks and Phom [Naga] leaders to resume their movement for independence etc., by giving them all sorts of false news. These tribes should therefore be kept under close supervision and guidance separately by keeping them engaged in various welfare and relief programmes.[23]

Regrouping was an advancement of territorializing space according to administrative convenience and political rule. The regrouping of villages was in progress at this stage and was an innovation of colonial strategies of governance. It was reminiscent of colonial policies, whereby British-administered villages were protected and assured of safety by punitive punishment carried out on unadministered villages.

By segregating villages and regrouping them, the frontier administration was reorganizing the whole social setup. At the same time, reorganization of land use was also taking shape with the introduction of wet-terrace rice cultivation that produced permanent rights over land that earlier were not fixed but circulated and shared by clan and *khel* members in a Naga village. This I shall discuss in Chapter 5.

People who collaborated with state officials like the *dobashis* and headmen were handsomely paid for their participation in state development projects. Naga rebels were seen as anti-nationals and were punished with forced labour, encounter killings and imprisonment. Villages were heavily guarded by the armed forces, and special provisions were made for villagers to be dependent on government supplies. During the counter-insurgency period of the 1950s, the grouping of villages brought loyal villagers close to the state. By providing systematic incentives and government supplies, the 'bad rebel' and 'good Naga' images were produced. The opening of *pani kheti* or wet-rice fields and economic 'welfare' incentives like free medical camps, the construction of new schools and driveable roads went hand in hand with military operations in remote villages, where it was claimed that hostiles received logistical support for their operations against the government forces. Any new settlements were looked upon with suspicion and were normally burned down. Man Bahadur Rai, in his 1957 tour report, wrote that these new sites where people could hide and conspire against the state were the remote places of the nation's frontier.[24] Rai's account states that he arrested the new settlers on the evidence that they had offered tea to the hostiles,[25] while a number of welfare measures were outlined for people who surrendered to government forces and were willing to be 'stockaded' in regrouped villages.[26] The panopticon of state control was underpinned by a national security incentive and the threat to the national identity by rebellious Naga people calling for self-determination.

Conclusion

In the first section of this chapter, I highlighted how state violence has been enacted by the Hobbesian understanding of order and disorder – the need to impose order over 'anarchy' – which needs to be demystified and challenged by a more humanistic and radical engagement with the practices of representing others and their cultural practices. This has been counter-posed at various points with narratives of violence produced by the people who are otherwise the objects of this state violence. However, far from producing a simple contrast of colonial/state oppressors and indigenous/local victims, the issue of the co-production of violence has emerged as an increasingly important theme and more so in the contemporary ceasefire environment. Violence is produced both by state agents and local elites who participate in state-sponsored development schemes. The struggle for stakeholdership in state-sponsored development programmes creates intense struggles for power, prestige and development money, which leads to scenarios such as the hostage scene as already discussed in Chapter 2.

In the colonial records and correspondence, we see the articulation of violence taking place. This was the discourse that defined violent acts as distinct from acts of administrative control. The discourse criminalizing headhunting, slave capture, raids and human sacrifice was as violent as those violent acts themselves. It is therefore the distinction between acts and discourse that needs to be studied carefully. Often, colonial officials were trapped in their schemes of ordering savagery by attacking villages that had neither committed raids nor were involved in any head-taking practices. As all information reached the colonial officials as communicated by their confidant informers from obedient villages, the colonial action often was based on the personal judgement of the colonial officials on duty. Depending so strongly on their relationship to political intermediaries, they did commit mistakes that resulted in heavy losses of life and property. For such lack of judgement there was no justice. Colonial authority was paramount. The tribal elders and tribal go-betweens were crucial in defining the direction of raids. This violence has to be understood not only as a form of knowledge produced through particular discourses privileged by particular officials, but also as an act of representation. The representations, like the acts, are co-produced by the various actors engaged in the sharing of power. As Zou (2005: 75) put it, 'headhunting is often neither the subject nor the object of study: it is a prolific site of discourse where the colonizer, the ethnographer, and even the local people engage in representing and translating the "other" as well as themselves, with diverse intentions'. To elaborate on Zou's remark in the context of Naga headhunting forays, colonial administrators who outlawed such practice as 'barbaric and hedonistic' since the late nineteenth century rewarded Naga headhunters who took enemy Japanese heads during World War II in the Naga Hills.

While interacting with Sanghpur village elders, I found at times that their narratives were distorted and unconvincing. Their representation of victims was narrated by people who were now in power. They translated the stories of raids to me. The colonial raids were organized and mediated by various patrons and go-betweens. This becomes very clear when we read colonial government proceedings that frequently acknowledge the help rendered. The colonial officials relied heavily upon the political go-betweens to exercise authority, mediate claims and resolve disputes. In tours that went forth with purely peaceful mediation in mind, the exercise of authority was more successful than those involved in punitive action. In the all-powerful village, actors shaped the colonial and local interactions. In the postcolonial period the number of go-betweens increased, with *dobashis* being a part of it. Today, political party workers, Members of the Legislative Assembly, Regional Council Members and formerly Area Council Members play the roles of intermediaries and benefactors in development projects and as negotiators of village–clan-level conflicts.

Issues raised in Chapter 2 now have more resonance. My own experience of being taken hostage and the events and understandings surrounding it connect with the narrative of the 1942 raid on Sanghpur organized by British official, Adams. Adams conducted the raid on false pretexts, according to the narrators. But if we look at the colonial tour diaries of people who visited this village, there was also awareness of instability and of violence with neighbouring villages. The British raiding party was then diverted by the *dobashis* to attack this village. Violence was thus co-produced. The representations are not conclusive; the historical context presents the village as being already hostile and at war with its neighbours. Similarly, the other experience, a personal one that I have elaborated as the story of my host being taken hostage by the Naga militants, reveals the postures taken by public leaders and how development programmes and the aid-dependence of the frontier produce the context for public postures taken by local public leaders to advance their interests and control resources. At the same time, these leaders are also deeply engaged in the vision of the state and act as collaborators of the state project. There is thus no neat divide between resistance and violence – or, indeed, between violence and 'normal' government and development processes. There are multiple agencies that become collaborators in the production of violence and similarly articulate their resistance to state power. The *jhum* villagers are caught in the crossfire. In the study village, the headmen narrated how they were tortured and mishandled by the underground people for not paying taxes and for leaking information on their movements, and that they were equally targeted by military patrols suspicious of their loyalty. Political parties, on the other hand, depended on them for their success in elections.

Such concerns played out more specifically with respect to my own position as an outsider having Aryan racial traits – always a worry among local student bodies and village institutions. However, initial suspicion lessened, and the sharing of information improved as time progressed. On the other hand, I also lived with a powerful village that had been associated with the local political party in power for two decades. This privilege opened opportunities as well as constrained me in approaching the people of the village.

I find it difficult to take a particular stance on violence. It seems at times directed from the state and at other times co-produced. To understand violence, I have used the historical records as primary text, anthropology and my own critical self-reflexivity as an ethnographer to narrate violence and also the construct of backwardness through my different experiences and reflections. For the colonial administrators, slavery, head-taking and feuding were barbaric and savage. For the native Naga subjects, state regulation and control produced violence and intimidation. In my own personal account, violence had different shades and meanings. At one point in time, I also experienced violence. For my analysis it is the discourse of violence and backwardness produced by different actors and how certain violent acts and modes of survival came to be powerfully articulated in 'public' as well as 'state transcripts' that must be considered more important than the violence itself. It is the knowledge and articulation of violence by different agents that matter. The colonial administrators were effectively representing slavery and head-taking as violence. Similarly, native actors in the contemporary context, the village intermediaries, saw certain acts as violence while negating others. For them corruption, poll-booth capturing and bootlegging do not produce violence. Violence is only produced by the hegemonic state, even if they at times collaborate with state planners and development officials.

Tsing writes in the context of Meratus mountains:

> The threat of violence operates at many levels. It is hard to understand the Meratus situation without thinking simultaneously about global

> projects of imperialism, the national military expansion that works with and against them, the regional administrative tactics that both frustrate and realize national plans, the local posturing of male minority group leaders' protecting their communities, and the ambivalent support and evasion of other community members who may alternately feel protected, threatened, or annoyed by the whole situation. (1996: 211)

In the study, village people's identity was often fluid. A person can become a church pastor, a foot soldier, a *musafir* (underground soldier), a village development board member or even a village council chairman in his lifetime. Genealogies and life histories produced no fixed identities or ideologies that could be strictly claimed to be anti-state or anarchist revolutionary. Past revolutionaries are deeply implicated in present state development projects, as they are the key beneficiaries and office bearers in the village. They were accused of corrupt practices. One of the most striking narratives was produced by a *musafir* himself. He was the son of my host's brother. In 1990 he travelled a long distance with the Naga army. Then one day he returned to the family farm. Challenged by his wife and three young children, he came to see farming as the only sustainable way to survive. He also works as an NNC comrade. In the village he is actively engaged in politics. In the mid-1990s, another faction from the Noklak area challenged him, and this made him abandon the faction he was recruited in. These days he is deeply involved in village political activities. He explained to me the hard work and uncertainties that befell Naga army soldiers and how difficult it was to keep a family while roaming around. In the village, he lives with his family and supports his father, who is a headman. He also collaborates in various development schemes and gives protection to political leaders during elections. He always promised that my presence in the village would never be questioned as people in the village knew that he was my friend. He had universal goodwill among his peer group. He also explained that he collected taxes from the villagers this season. The village non-tribal shopkeeper was moderately taxed at Rs 500, as they were on good terms. He called this reciprocity as *moram* (mercy).

This word I heard used several times by my host when we were taken hostage. Many people in the village also used the term when they were in financial crisis. When somebody came and asked for money, they uttered this word. This was while he explained that he made a compromise with his cousins: the rice mill was taxed just Rs 100. Here the word *moram* took a separate meaning: it was likened to blood relationship, generosity, and acts of kindness for favours in return. By being assigned as the tax collector, he exercises authority over common villagers and mediates terms of relations. His narratives of good and evil varied from a common villager or a political party leader who understood violence differently and interpreted it differently. Violence is at best co-produced in this context by state incentives, policies and the individual agency of local leaders who exercise their authority. The diffusion of democracy is a postcolonial phenomenon in the hills of Northeast India as democratic decentralization and community-driven development has been a recent phenomenon (after years of centralized planning under NEFA administration and military intervention in the borderlands). Democracy itself has produced new forms of patronage that link Naga villagers with the state and reproduce violence through party politics, calculative matchmaking, bribing and the buying of votes during elections. Community feasts and distribution of free alcohol during elections produce a messy electoral environment in which people who resisted the party protocols laid down by political intermediaries faced execution and threats from village henchman.

My position in the field was similar; I was caught up in the violence. Violence in my study area is co-produced by the ruler and the ruled. There are different narratives for violence that demand explanation. My involvement in the field is personal and experiential. Violence in the postcolonial context is surreptitiously embedded in democracy through the circulation of cash and patronage-building with the community clan, lineages and party workers in the villages, where resources as well as arms are mobilized to buy votes. The earlier forms of control through military tours, gift-giving and *dobashis'* central role in mediating claims are now replaced by government staff, district administrators, political party workers, village development board volunteers, village-level workers and village council chairmen, who link the people with the administration and the state through improvement schemes and state projects. The decline of the *dobashis'* power matches the rise of political leaders who are former *dobashis,* and most are political actors. These days *dobashis* enjoy fixed salaries and have been pushed to helping the administration and politicians.

The different forms of administration, as commonly depicted as colonial, post-independence frontier rule and, of late, decentralized governance based on popular participation, have produced distinct forms of control over the Nagas. So have the colonial and missionary practices of civilizing the Nagas from the practices of head-taking and human sacrifice. However, innovations and reforms in administrative control did not disrupt the character of this place as being marginal and a site of greater state authority and control. Instead, state intervention and control over local people intensified, with developmental grants and networks of programmes that now link the local with the global. The links are established through local everyday politics, talks on conservation of wildlife and millennium goals of sustainable livelihood through community-driven participatory programmes that include as well as exclude people from the sources of development. The forms of violence that are produced through improvement schemes are rarely talked about and are a 'taboo' as the anthropology of the Nagas has been silent on the work of politics and administration. Similarly, violence has attracted scanty attention. Violence is only seen as being produced out of political subjugation of self-determination or the quelling of ethno-political rebellion and the long-standing political demand and resistance of the Nagas. Other forms of violence are rarely talked about in academia and narrated in historical or even ethnographic accounts. Violence has attained a stereotype associated with standard 'European' notions that sharply define them as acts that distinguish barbarity from civilization. In the activist mind, violence is that which is physical and not that which is innate in an era of ceasefires and negotiations.

For the Naga village where I conducted fieldwork, a colonial and postcolonial 'out-of-the-way place', the no-man's land of the Indo–Burma frontier, violence is omnipresent and is expressed through 'administration', 'pacification', 'panoptic vigilance', and the new partner 'development', continued in the name of quelling savagery and backwardness. Development has emerged as the new 'avatar' to which the aspiration of the people's improvement has been attached. Nobody wants to live in the village, perform shifting cultivation and do the kind of activities that the last generation did. People are looking for 'improvement'. Often improvement comes with a cost, as it brings about disruptions to traditional practices. But out of these heterogeneous struggles, people aspire for development and improvement in their own lives. The next chapter turns to discussing state–society relations – understanding how *jhum* cultivators were tackled by the colonial administration in the frontier through political go-betweens and village intermediaries who acted as trustees to the colonial state and the guardians of Naga villages. I will also discuss how the agrarian frontier evolved with the promotion of wet-terrace rice agriculture as a technology to settle the *jhummias*.

4

Jhum and the 'Science of Empire': Ecological Discourse, Ethnographic Knowledge and Colonial Mediation

This chapter examines colonial and postcolonial depictions of *jhum* cultivation in the Naga Hills and the policies (sporadically) pursued. It teases out the tensions distinctive to this region as the political and administrative frontier of British India. In the nineteenth century, the discovery of tea, minerals, notably coal and petroleum, and the expansion of the Assam–Bengal railway along the Naga foothills adjoining the Assam plains (in the Sibsagar District) added to the complexity of fixing the boundaries between the hills and the plains as the 'tea frontier' and the 'Inner Line' – the political and administrative divide between the two worlds, hills and plains – edged further into the hills. In 1873 the arrival of a new player, the Forest Department, added further issues. The colonial administration was constantly confronting local settlers, both Naga and non-Naga, who by now had become part of the tea frontier. Beyond the 'Inner Line' boundary, in the territory of the administered Naga Hills District, the local colonial administration devised policies of 'least interference' – or indirect intervention – by which a thinly administered area could be effectively managed by interacting with Naga customs and traditions (as understood by the administration). This chapter concentrates on explaining how this broad political stance was decisive in formulating colonial policy on *jhum* cultivation, not only in the nineteenth century, but also in the twentieth.

'Non-interference' did not mean there were no interventions with respect to *jhum*. On the contrary, in the name of 'progress' and 'improvement', colonial officials initiated a number of schemes. However, the crucial point is that *jhum* policy in the region was above all dictated by political–administrative concerns determined by the local exigencies of rule: to establish political control through 'moderate' forms of political administration. The patronage system established by hill officers with village headmen and chiefs was highly important (see also Chapter 5), as were attempts to transfer existing local methods of cultivation between tribes, which in turn depended on the approval of and adoption by the local headmen and chiefs. In this way, local administrators followed a policy of adopting and replicating local 'best practice' for *jhum* 'improvement'. Drastic change was seen as detrimental to effective administration of the Naga way of life in this politically contested frontier. Thus, however, negative were the general views held by local administrators (or, indeed, forest officers) about *jhum*, it was accepted as a given that no radical measures could be adopted in the Naga Hills; such was the predicament of control.

The *jhum* policies adopted in the Naga Hills were thus quite different from other upland areas of British India, where *jhum* was branded a destructive and, in modern terms, unsustainable form of agriculture that devours the forest. Notoriously, in those areas the *jhummias* (*jhum* cultivators) were systematically forced to change their farming practices and thus were often isolated from their wider networks of livelihoods outside the *jhum*-forest world (see Elwin 1939; Prasad 1994; Rangarajan 1996; Sivaramakrishnan 1999; Bryant 1997). In these other parts of the colony, generating tax, reserving the forest for scientific regeneration and other revenues played a key part in the forcible imposition of settled agriculture. By contrast, in the Naga Hills revenue per se was one of the least important concerns of the administration. The collection of a 'hoe tax' or house tax from *jhum* cultivators in the administered areas was seen, rather, as an instrument of rule – a way to keep a record of the semi-nomadic Nagas, making the mobile *jhum* space 'legible' (Scott 1998). In the same vein, agricultural interventions and sedentarized forms of agriculture may be seen as primarily driven by the desire to produce a governable space with a settled population rather than as revenue driven.

In this context, the colonial science debates on soil desiccation and erosion that came to surround shifting cultivation were not paramount in determining colonial policy making, even though by the early 1900s, rain-gauge stations were established in Kohima and Mokokchung subdivision, and annual reports of the Naga Hills soon after began to record monthly average rainfall and temperature. Rather, reflecting the local political exigencies of rule, what dominated colonial perceptions of *jhum* in the Naga Hills were the 'cultural technologies of rule' (Cohn 1996) or what Gilmartin (1994) calls the 'science of empire' in contrast to 'imperial science' (to which I shall return below). *Jhum* was understood by local colonial administrators in terms of the wider tribal ethnography of the 'primitive'. Shifting cultivation, regarded as the most primordial form of crop production, was seen as wasteful, even savage and wanton. Once permanent control had been established over the savage Nagas in the administered areas, interventions around *jhum* were conceived as bringing civility and order through 'improvement'. In turn, reflecting a scale of degrees of civilization, or the even earlier distinctions between noble and ignoble savages, attempts were made to disseminate the wet-terrace rice practices of the more 'advanced' tribes (the Angami) to other Naga tribes. While respecting indigenous local practice, *jhum* could thus be gradually modified and, ideally, replaced by more productive and sedentarized forms of farming that were closer to civilized settled forms of livelihood.

However, scientific approaches, in recording climatic data, did play a role in selecting cash crops that were introduced in the Naga Hills. For example, in the early decades of the twentieth century the daily record of rainfall and temperature at the field stations across the hills generated valuable data that helped the Agriculture Department shape policies for promoting sustainable agriculture in the hills. Meanwhile, in the 1920s and particularly in the 1930s, forestry officials, including in Assam, were heard registering their concerns about the dangers of *jhum*. But the administrators remained largely committed to their own politically defined concerns. Even where this involved taking a position 'contrary to the views expressed by leading foresters in the twentieth century for stricter control over shifting cultivators, the government had no option but to hold a very ambiguous position regarding the practice [of *jhum*] even though they were no less worried by the destruction caused' (Saikia 2011: 274). In the end what emerged was a compromise – the 1946 *Jhum* Land Regulation.

This chapter also examines the policies adopted just after World War II and on the eve of Indian independence, especially the 'Grow More Food Campaign'. This latter nationwide campaign to grow more food was pursued by the Imperial Agricultural Department in light of the famines and food shortages that were partly caused by population pressure on land. The Grow More Food programme, as Ross (2003) argues, presented India as a 'Malthusian case'. Although, the Grow More Food programme came to an end with India's Independence, the new five-year plan initiatives such as the Community Development Programme (CDP) and the Integrated Area Development Programme (IADP) came to carry forward the premise of neo-Malthusian population-induced environmental change. In the Naga Hills, however, the political exigencies had become a matter of internal security, a threat to India's sovereignty post-1947, as the vast majority of the Naga people declined joining the Indian union and instead chose the path of armed struggle. The post-independence government found it legitimate to nationalize frontier spaces through colonial strategies of rule – indirect control and the promotion of local measures to sedentarize farming (these were coupled with calculated counter-insurgency campaigns involving the strategic hamleting of Naga settlements that would lead to the defeat of insurgents) (Rai 1956; Sundar 2011). This, I argue, is a significant departure from the agricultural-improvement policies followed in the rest of India. The seeds of the Green Revolution never touched the Naga Hills. In the late 1990s, this project of sedentarization took a new turn with a shift in emphasis from wet-rice intensification programmes to upland agri-silvicultural development through community development programmes and through the decentralization of governance.[1] This is discussed in Chapter 5, where I examine the changing values attached to land and the developments in crop cultivation and changing land use.

In the end, both colonial and postcolonial attempts were largely unsuccessful in making a significant impact on transforming Naga farming. The problems associated with implementing such policies were huge, and the colonial administration could not sustain rice-intensification programmes on a large scale or over long periods of time. The farmers who were introduced to the new crops and land use did not see any major value in such farming. Farmers quickly reverted to opening new *jhum* plots once the incentives to grow terrace rice dried up.[2]

Regulating Frontier Spaces: Tea and the Forest Frontier

In 1840, British aspirations to compete with China's tea industry inspired Robert Bruce, a colonial tea planter stationed in Assam, to 'discover' local tea in the province, cultivated in the wild by the Singphos tribe in the Upper Assam Valley. Subsequently, the colonial government (at that point the East India Company) attempted to colonize the Upper Assam Valley with tea plantations by giving 'wasteland grants' to British tea planters and rubber speculators to clear dense forests not suitable for agriculture (Sharma 2006, 2010). The virgin forest that remained unused by the Assam peasantry or periodically cleared for *jhum* was systematically settled through the Wasteland Grant Act passed in 1843 (preceded by the 1838 Waste Land Grant Rules). This legislation gave powers to the local administration to grant land in favour of mining syndicates and British tea planters, who took up these lands for the expansion of their plantation farms and for permanently converting the 'wasteland' to plantations and mining lands.

Pressure increased as virgin forest was further demarcated to open railway lines for the Assam–Bengal Railway that was central to the export of tea, coal and petroleum discovered in Upper Assam during this period and on which work started in the late nineteenth century (Barpujari 2003; Saikia 2011). These policies were pursued with a new resource regime in mind, one that needed a number of legislative and administrative changes to control the plains and hill populations and set a new land tenure and propriety regime in Assam (see Baruah 2001), thus opening up the frontier to colonial capital and intensive land regulation. A chronological citation of the legislation passed for hill districts and Assam proper explains the situation (see Table 4.1).

Table 4.1 Chronology of legislation passed in Assam since 1834 by the British Indian Administration

Assam Frontier Tracts Regulation, 1880 (Reg II of 1880)	
1834 Assam Land Act	
1838 Assam Waste Land Grant Rules	
1847 Scheduled District Act (covering the whole of Assam province)	Enabled the executive to exclude 'Scheduled District' areas from the normal operation of any ordinary law. The Act declared the entire province of Assam a Scheduled District. Enforced in 1880, Frontier Tract Regulation Act. *
1866–67 Naga Hills District created as an 'administered area' (notified January 1867)	Although boundaries unclear
1868 Forest Revenue System created	A new revenue system under which small *Tehsildars*, called *Mouzadars* were appointed, charged with protecting forestland and collecting tax revenue (*Ryotwari* and *Zamindari* System – both operated in mix in Assam).
1873 'Inner Line Regulation'	Bengal Eastern Frontier Regulation Act (1873) made it lawful for the state government to prescribe (and move) the 'Inner Line' in the districts of Kamrup, Darrang, Nowgong, Sibsagar, Lakhimpur (Garo Hills), Khasi and Jaintia Hills, Naga Hills and Cachar. *
1873 creation of Assam Forest Department (previously 1868 under East Bengal)	
1884 Agriculturalists Loan Act	
1886 Assam Land and Revenue Regulation (Regulation I of 1886, did not apply to Naga Hills)	
Regulation II of 1889 came into force on the 21 September 1889	
1891 Assam Forest Regulations	
Regulation II of 1905 c Part 7 Hoe Tax or House Tax	
1911 Chin Hills Regulations (1896)	Extended to the Naga Hills on 9 October 1911
1927 Forest Act	
1946 *Jhum* Land Regulations	

**Source*: Upadhya and Jain (2004).

The competition for wastelands intensified when the Imperial Forest Department appeared on the scene. The department was created in 1864, appearing in Assam in 1868 as part of the East Bengal Forest Department (at this point Assam was part of East Bengal) and finally was established in its own right in Assam in the 1870s. The Forest Department was a friend neither to the plains nor the hills people – that is, neither to (state-promoted) agrarian plantation agriculture nor to *jhum* cultivation – seeing both as threats to the forest.

Conflicts over farms (major tea planters) and forest became more acute (Saikia 2011), especially in areas bordering on the foothills. The colonial hills administration and the deputy commissioner had the power to act as the arbitrator of disputes that emerged between the *jhum* settlements in the hills and the tea cultivators on the plains. To protect tea planters and the coolies on the plantations from Naga raids, the government sent many punitive expeditions to the Naga areas until the late 1860s, when the Naga Hills began to be permanently administered with the creation of a district (Allen 1905; Ao 1970). The establishment of the Naga Hills District had been followed by the Inner Line Regulation of 1873, which restricted the access of the plains population to the hills through a pass system and set the territorial limits of revenue administration in the hills. From 1873 to 1930, the Inner Line was modified several times, largely influenced by the tea planters and other commercial interests, thus facilitating the expansion of tea gardens, mining, and so forth into the Naga foothills. At the same time the Forest Department also sought to carve out valuable forestland from the Naga Hills. The Forest Department was in competition with the tea plantations:

> Most tea plantations were established by clearing natural forests on lands acquired or purchased from the Government. Whenever markets for tea were strong enough to enable expansion of plantation acreage, forest cover correspondingly got reduced. For many years the tea planters held dominant financial and political leverage in Assam, and prevented the Forest Department (their competitor for control of forest lands) from gaining control over wide forest areas. (Handique 2004: 42)

Across India, the Forest Department has been characterized as pro-conservancy in its early years, turning to more purely revenue-oriented interests towards the end of the nineteenth century. However, in this period 'it is important to note that the British account of the forestry situation in the northeast is confined to forest production plans, and even conservancy was primarily aimed at securing a permanent supply of timber through a state monopoly' (Upadhya and Jain, 2004: 3). In Assam, the creation of reserve forest was initially focused on securing a permanent supply of timber through state monopoly in order to provide sleepers for the Assam–Bengal railway line and to supply wood for tea crates for the ever-expanding tea plantations (Saikia 2011: 267).

In terms of property rights, crudely put, the Forest Acts declared all 'reserve forests' as the property of the state. A lesser right (in district forests) gave the Forest Department command over certain reserved standing trees (as notified periodically in the *State Gazette*), although not the land they grew on.

> State monopoly over forests was first asserted through the Indian Forest Act of 1865. This law simply established the government's claims over forests. Thirteen years later, however, it was reissued with far-reaching amendments as the Forest Act of 1878. This version curtailed centuries-old, customary-use rights of local communities over forests and consolidated the government's control over all forests. The Forest Act of 1878 established that forest use by villagers was not a right but a privilege of concession given by the government. (Mitra and Gupta 2009: 195)

The creation of reserve forests in the administered areas brought in enormous conflict with the Naga shifting cultivators as their right of access over forests was abrogated and their access to forest resources denied. Most critical was the potential effect on *jhum* land within the state forests. Even where *jhumming* was allowed within the forests (through a permit system), under the Assam Forest Regulation 1891, this was legally in the form of a privilege rather than a right.[3] Within this context, in its First Annual Forest Report the Assam Forest Department conceded that *jhum* cultivation 'must be permitted' as 'the people lived by *jhumming*, a stoppage of which would mean starvation to the people if it were attempted and is not advisable for political reasons either' (quoted in Upadhya and Jain 2004: 3). Indeed, Saikia notes that, in the early days, foresters 'had no clear views on shifting cultivation and whether it caused deterioration of forest cover' (2011: 273).

A nineteenth-century dispute illustrates the interlinkage of Inner Line modifications and the creation of forest reserves. In 1884 the Inner Line was moved in order to facilitate the mining operations of the Assam Railway and Trading Company and, as a consequence, six Naga villages in the Jhanzi area of the foothills in the Lakhimpur district now fell outside the Inner Line and hence within the general administration of the government of India and, at the same time, the Jhanzi forest was declared a reserve forest. Tensions surfaced as the forest officials who had entered the Jhanzi reserve forest met with resistance from Naga villagers in the winter of 1884. It was later realized that the notification of the modified boundary, issued on 16 October 1883 could never have been communicated to the Naga inhabitants, as the proclamation was served only to *Mauzadars* in the plains. No further steps had been taken by the district officials to give intimation to the Nagas, or to invite a statement of their claims. In the following months it became clear from increasing resistance posed by Naga villagers that the local Nagas had a strong claim. The Jhanzi reserve forest was declared a district forest, thus restoring the rights of the Naga villagers to access the forest resources,[4] to collect firewood and other minor forest products. As Saikia explains, this involved some juggling: 'In this case Mann, the Conservator, thought it advisable to throw open the reserve and fill the gap by reserving land in adjoining districts' (2011: 273).

A number of such disputes were recorded from the last decades of the nineteenth century as the reserve demarcation became a common ground for disquiet between the foothills Naga settlements and the colonial Forest Department, with the department now taking control over the forests and settling these lands as reserves and forest villages[5] and as district forests. Saikia also mentions another case, in 1879 in the Mikir Hills, where the political officer and forest authorities came into conflict. The political officer initially refused the application from the Forest Department for extension of forest reserves, on the grounds that this 'would put local people in "dire straits" [in gaining access to] fresh land for *jhum* cultivation'. The Conservator overruled him on the grounds that 'the area was sparsely populated' (2011: 273). The economic potential of frontier forest was disturbed by native inhabitants who lived in nucleated settlements sparsely distributed over the vast forest landscape. However, according to Saikia, the Forest Department came around to the dominant political view: 'gradually another idea came to take shape among foresters and they suggested that change in this mode of cultivation could only be brought about very gradually and cautiously' (2011: 273). It was not until the early years of the twentieth century that the forestry officers began to oppose *jhum* on scientific conservancy grounds, thus linking economic potential and scientific practices of forest management. Forests came to be seen as both 'economic goods and as laboratories of scientific and epistemic enterprise', combining regenerative forest activity, with scientific conservancy to create wealth. This can be accessed from the observation made by a colonial official in Assam in the late nineteenth century:

> It is hard in my province to more than allude to the noble forest covering the whole country save where the Nagas and other tribes have cleared a space on the hills or where in the plains a small patch of cultivation supporting a Cachar village appears like a small speck, an infinitesimal island, in the vast sea of jungle stretching further than the eye can reach. These forests seem destined to remain for ages as they now are, and to be a vast storehouse of timber wealth to meet the ever increasing wants

> of civilization; and at the same time we must not be too prodigal of our wealth, and no one who has not had experience of forestry can for a moment understand how small is the proportion of trees useful for building, compared to those useless except as fuel.[6]

By the 1930s the colonial forest administration had established ten forest reserves along the Naga foothills.[7] Similarly, a substantial portion of the virgin forest in the foothills near Dimapur, Amguri, Simulguri in Golaghat and Sibsagar district had been brought under state forest management through wasteland grant and were even declared reserves. The reserve forests excluded the Naga *jhum* cultivators and marginalized their livelihood by permanently restricting their entry into these reserves. In the forest reserves, Nepali woodcutters were resettled for various regenerative and scientific forestry works through working plans prepared by the Forest Department. The settled villagers were provided with land to grow paddy in formerly *jhum* fallow occupied by the Forest Department through the payment of paltry compensation to Naga chiefs who claimed hereditary clan rights to *jhum* the land. This resulted in violent conflicts all along the forest frontier, often resulting in destruction of new plantations, crop raiding by the neighbouring Naga villages and claims over the produce of the new forest inhabitants (immigrant Nepali planters/forest villagers and Adivasis, who had come as an indentured labour force to work on the plantations).

Coinciding with these policies of land appropriation by the Forest Department for the creation of reserve forest, the Chin Hills Regulation of 1896 was applied to the Naga Hills in 1911. (The Chin Hills themselves were in 'Upper Burma' in 1896). Chapter VI of the regulation dealt with taxes and fines. Section 35 enabled local government to levy taxes 'on all clans and villages at such rates and in such manner as the Local Government may prescribe'. For example, the administration classified land in Dimapur *mauza* (which lay in the plains bordering the Assam–Naga Hills) into five categories, which were assessed differently.[8] Agricultural land was taxed at different rates based on its quality (productivity). For non-*pattadars* (i.e., people without land deeds) – people who primarily did *jhum* or were labourers – land assessments were at a much lower rate, collected on a household basis (house tax or hoe tax) as they occupied forestland.[9] According to colonial revenue officials, they were poor people occupying non-productive wastelands. In a 1939 memorandum J. P. Mills, by then secretary to the Governor of Assam, writes that 'the minimum land revenue shall be three rupees per annum for indigenous people and five rupees per annum for foreigners[10] (by 'foreigners' he meant non-Nagas who had settled in the mauza and were landholders).

Jhum Land Regulation 1946

In 1946 the Assam administration, under section 92 of the Government of India Act 1935, introduced Regulation III of 1946, known as the Naga Hills *Jhum* Land Regulation. On the one hand, it reflected the long-standing government policy of minimal interference, and indeed strengthened that policy by recognizing significant 'traditional' rights over *jhum* land. On the other hand, it registered growing pressure for conservation. Yet what might have looked like a protection of *jhummias* radically altered the shape of landholding in ways that would have disastrous consequences for land use and land relations (see chapter 5).

The regulation recognized 'customary rights' over shifting cultivation land in two categories (a) community *jhum* land (where there were no fixed plots) and (b) individual *jhum* land. The first category section, hereafter Section 4(1), established such rights 'in favour of a village or community when such village or community has enjoyed the right to cultivate or utilize such *jhum* land for not less than 5 years prior to the making of this regulation'. The second category, Section 4(2) stated:

> A customary right to '*jhum*' land shall be deemed to be established in favour of an individual cultivator if:
> (a) He has inherited the land in accordance with the local custom.
> (b) He purchased the land prior to the making of this regulation and such purchase was not contrary to the local custom.
> (c) He has purchased the land at any date subsequent to the making of this regulation, provided such purchase was not contrary to the local custom.
> (d) Being a resident of a permanent village, he has brought the land under cultivation, and the land has not been cultivated at any time within 30 years preceding his bringing the same into cultivation provided such land is within cultivable reach of his own village.

The regulation further provided restrictions on the transfer or leasing of *jhum* land to communities or villages or individuals other than those with the customary right, although Section 5(2) permitted transfer of *jhum* land to which an individual had customary rights: 'To another member of the same village or community or to that village or community as a whole.' Section 6 specified that 'local people having customary rights to a *jhum* land shall be entitled to forest produce for bona fide use'.

On the face of it, this appears as a progressive legislation. Where *jhumming* was recognized only as a privilege under the Assam Forest Regulation

1891, the recognition of customary rights gave communities/villages and individuals permanent, heritable and transferable rights.[11] Equally, such rights were not to be used to alienate land, temporarily or permanently to other villages/communities. However, there were a number of problems.

First, the provisions were carefully designed to settle the *jhummias* in villages and designated territories. In order to claim rights, a village or community had to be settled in one location (although migration to *jhum* sites for part of the year was not a disqualification). Section 2(b) defined *jhum* land as 'all land which any member of a village or community has a customary right to cultivate by means of shifting cultivation or to utilize by clearing jungle or grazing livestock *provided that such a village or community is in a permanent location*' (emphasis added). Section 3(2) specified further that 'any village or community shall be held to be in a permanent location if it always remains within a specific area, although part or the whole of such village may migrate from time to time to different localities within that area'. Equally, for an individual to claim rights, there had to be a member of a 'permanent village'. In other words, the emphasis was on the existence of a permanent settlement.

Second, in establishing permanent tenureship rights, the ground was laid for the emergence of full private property rights (i.e., alienable by lease or sale without restriction). Since the provisions allowed *jhummias* to establish their rights permanently over common lands that remained fallow for more than 30 years, this was especially significant. These provisions allowed the growth of private tenure in villages where wet-rice cultivation was intensified in the 1950s (see Chapter 5).

Meanwhile, the regulation sought to control land alienation by specifying that no such transfers or leases outside the village or community were permitted – except with the permission of the land conservator, whose role was to be assumed by the deputy commissioner and/or his subdivisional officials, who had full powers to arbitrate any dispute related to land and forest (Section 2(a)). Section 9 specified that *jhum* land was taxable, and that failure to pay such tax would allow the deputy commissioner (DC) to 'declare such rights of that person to have been extinguished' and to transfer the land as he saw fit. These powers in the hands of the DC were in turn passed on to other forms of local administration, ultimately the village boards. Thus, the *Jhum* Land Regulation of 1946 was an important weapon in the hands of the local administrators, as the government was given the upper hand in controlling the lives of the native *jhum* cultivators.

It is true that Section 15 specified that where an approved tribal council had been set up in any area, 'Any or all of the powers granted to the Land Conservator under the provisions of this regulation may be vested in that Tribal Council in respect of such area and the Land Conservator shall not exercise any power so vested.' However, this vesting of powers was itself to have ill effects, as rivalries and factionalism grew within them.

Jhum, Scientific Knowledge and Empire Building

By the late 1930s colonial governments all over the empire were discussing the impact of shifting cultivation on soil erosion and soil desiccation. Erosion and desiccation narratives became particularly vocal in the colonial discourse because both soil and forest were crucial for colonial revenue. In this context, swidden farming was interpreted by colonial science[12] – or 'imperial science' – as a 'crisis' across the colonies (Rajan 1998).

The idea and image of conservation had already been in circulation in the Colonial Agricultural Conference (CAC) (1920–50), and in the Colonial Office for some time (Grove 1995). Leading proponents of conservation, such as Baden Powell in the 1870s, had been propagating the image of swidden farming as wasteful and harmful. The linking of *jhumming* to deforestation, forest fires, floods and changes in global climatic conditions had taken shape during this early period but it was not until later, in the 1920s and 1930s that soil erosion and desiccation became leading concerns.

In the Naga Hills by contrast, even where local administrators did promote settled farming over *jhum*, this was more likely to have been due to the lack of prospects for revenue generation and the problems associated with the collection of tax from nomadic *jhummias*. John Rowntree, a forest officer in Assam in the 1940s, records the frustrations of trying to deal with the problem of *jhum*, as perceived from the Forestry Department perspective:

> *Jhumming* is one of the most serious problems in the hills of India and one which we [the Forest Department], in our time, failed to solve, despite many attempts to do so. In the reserve forest[13] villagers were allowed to *jhum* in unproductive areas, on the condition that they planted the land with forest once they moved on, a practice known as *taungya*, but this hardly touched the fringe of the problem. In a few places, they were persuaded to adopt a system of permanent terraced cultivation, but the hill men are a conservative lot, reluctant to change their ways, and *the politicians took a lot of persuasion that they should do so.* (Rowntree 1987: 12, emphasis added)

This picture of local forestry officials (and local agricultural departments) trying to get their conservationist concerns registered is familiar from contemporary debates about 'colonial/imperial science' in the nineteenth and twentieth centuries. At the risk of oversimplification, these debates – and associated periodizations – may be summarized as follows. The 'centralist position' sees the metropolitan 'centre' (the CAC) as imposing its Eurocentric 'high science' on the colonies. The 'localist' response is that, in fact, much of the substance of these scientific perspectives came from 'below' rather than being imposed from 'above'. It was local, relatively mid-level, colonial officials (notably forest officials) – relying on 'low technology' (such as the routine establishment of surveys and meteorological stations recording climate factors) rather than 'high' theoretical analysis oriented to European agriculture – that produced the soil erosion and soil desiccation conservationist discourse on swidden agriculture. This soil erosion and soil desiccation concerns emerged relatively early at the local level but only began to register in central/imperial debates in the 1920s, largely focused not on *jhum* but on the negative effects of introduced cash crops in the Caribbean and East Africa (Hodge 2007: 144). From this perspective, the issue is how long it took for 'the concerns of local authorities [to be] lifted onto the imperial stage' (Hodge 2007: 153), the 1930s being the high point.

Both sides of the debate appear, however, to agree on the preoccupations of 'imperial/colonial science' before the 1930s as being driven by a commodity- and revenue-based research agenda. According to Hodge, the 'introduction and improvement of cash crops such as sugar cane, cotton, rubber, coffee, palm oil, and tea' were high on the agenda, with scientists focused on applying genetic research on 'higher-yielding and more drought-resistant crop varieties' and the 'eradication of insect pests and plant diseases' – all 'heavily biased in favour of plantation enterprises and settler estate farming'. While, 'in areas of peasant cultivation, preference was given to the introduction of "sound" European agricultural practices' in order to improve yields', also introducing 'exotic crops rather than local varieties' (Hodge 2007: 147–48). Simultaneously, tropical forests were approached in the same spirit, focusing on how to extract maximum benefits, whether in the form of usable resources or revenue. All this was predicated on a fantasy of infinite abundance, because it was never imagined that the vast fertility of tropical forest resources could ever run out. Thus, the first appearance of proto-ecological or conservationist concerns was in respect to the damage that British exploitation of natural resources was causing. However, these concerns were then 'dumped' on *jhumming*.

To what extent does this picture apply to *jhum* policy in the Naga Hills? Some parts of the picture fit. In Assam, forests supplied the raw materials for the expansion of the tea trade (wood for tea crates) and for the railway (wood for railway sleepers). There were also attempts to introduce cash crops (see next section). In Assam, rain gauge stations had been established in the Kohima and Mokokchung subdivision in the early 1900s and annual reports of the Naga Hills soon started to report the monthly averages of rainfall and temperature. As noted earlier, these were indeed critical to the introduction of cash crops and the choice of what to introduce where.

Rajan (1998) argues that the Naga Hills were not entirely segregated from such concerns to the extent that *jhum* practices in the Naga Hills were also considered from the point of view of their indirect impact on the plains. Shifting cultivation in this regard provided a triple threat to colonial revenue prospects in the Indian subcontinent: firstly, through deforestation of valuable timber forest and, secondly, through topsoil erosion induced by burning in the hills. Thirdly, transportation of high sediment load by rivers aggravated the flooding situation in the lower plains. In Assam and Bengal this became particularly important as the plains areas, unlike the hills, were positive avenues of revenue, while the hills were areas of deficit. Thus, it was of utmost interest for colonial officials to protect the hills in the Eastern Himalayas from being denuded and to conserve its forest so as to minimize the hazard posed to the Brahmaputra and Ganges flood plains.

However, the terms of this debate cannot be simply transposed onto the Naga Hills at large. While some minor revenue-creating activities were introduced, the region, including the forests, was never seen as a major source of revenue (as discussed below). To a large extent these concerns were – at least until the 1930s and 1940s – governed by an entirely different sort of discourse, in which *jhum* was understood not so much in terms of imperial/colonial science but through a different 'science', more oriented to the social sciences than the natural sciences: the 'science of empire' with its focus on to the pragmatic processes of rule and the administrative language of the colonial state.

Savagery and Agriculture: 'Imperial Science' versus the 'Science of Empire'

While colonial conservation policy was informed by the empire's scientific conferences on soil desiccation and climate change, the effects of deforestation and soil erosion caused by hill farming – primarily swidden and grazing – carried out by hill cultivators in the upland areas was influenced at the local level by colonial official perceptions of *jhum* and observations made by colonial administrator anthropologists, survey parties, forest officers and later forest conservators and agricultural officers in their reports and correspondence on administration and rule.[14]

Long before twentieth century 'scientific' concerns came into play, local administrators in the Naga Hills were describing and making pronouncements on *jhum* practices in their tour reports. This administrative–ethnographic discourse can be contrasted to natural science/imperial science approaches, whether 'central' or 'local', and characterized, rather, in terms of what Gilmartin (1994) calls the 'science of empire':

> The imperatives of 'imperial science, the worldwide application of science to the productive control of nature,[15] were closely associated with the expansion of colonialism and commercial agriculture. But the transformative operation of 'imperial science' was constrained in the colonial context by the presence of another powerful scientific 'discourse' of power, the 'science of empire,' or imperial administration. (Gilmartin 1994: 1127)

The 'science of empire', in contrast to 'imperial science', refers to precisely the sort of regularized ethnographic recording already described as forming a key part of colonial administrators' routine practice in annual tour diaries and even punitive expeditions and, in their published monographs, an early ethnographic discourse that 'incorporated idioms of custom and genealogy into a new legal and administrative language of the colonial state' (Gilmartin 1994: 1127). As Mitra and Gupta argue, rather than seeing *jhum* as just an agricultural practice it was seen as a manifestation of the nature of tribes:

> The stereotypical attitudes of colonial administrators toward tribal uses of forests were most acutely captured in colonial policy on shifting cultivation, or *jhum* agriculture. Shifting cultivation, also known as swidden, is not an exclusively tribal practice, but it characterizes tribes, as it is different from the dominant culture of the plough that has been a key characteristic of the mainstream Hindu society. (Mitra and Gupta 2009: 196)

Gilmartin also emphasizes the connection between such 'local knowledge' and systems of indirect rule:

> Most critical to the 'science of empire,' however, was the fact that it defined not simply a structure of domination, but also a distinctly colonial political system linking the colonial state and indigenous elites together in a common political order. The key to this order and the 'science of empire' lay in the scientific appropriation of what might be called 'local knowledge.' Indigenous 'local knowledge' (and the local power relations and forms of knowing that it embodied) of necessity occupied a critical place in the scientific structuring of colonial administration. (Gilmartin 1994: 1128)

The administrative–ethnographic discourse was a crucial 'cultural technology of rule' (Cohn 1996). For the 'administrator–ethnographers' the inhabitants of the Naga Hills represented primitive societies almost untouched by (Western or Hindu) civilization. *Jhum* cultivation with its associated nomadic mobility was a key part of this picture. It was this local discourse, rather than either 'central' imperial science or the interventions of local foresters and agricultural department officials, that was the dominant determinant of colonial policy on *jhum*. This approach located *jhum* cultivation, on the one hand, in terms of a scale of primitiveness–civilization and, on the other hand, in terms of the idea of 'wastefulness', which owes more to eighteenth century concerns than to nineteenth-century science (Hodge 2007).

The colonial ethnographic monographs on the Semas, Rengmas, Aos, Angamis, Lothas and Thangkhuls were aimed at informing the colonial officials on duty about native life and state of affairs and to rescue and record their customs before they disappeared. They give detailed accounts of crops cultivated, the *gennas* (taboos) and customs associated with *jhum* cultivation. The elaborate descriptions are aimed not only at a wider scholarly audience but at informing the local administration about native life and its subsistence practices. The descriptions of rituals associated with *jhum* farming represent a static way of life. While the avowed policy of the colonial state was one of 'least interference', such mapping of native customs, practices and rituals was undertaken in order to maintain control and better administration.

At the same time, this seemingly dispassionate mode of description was contiguous with depictions of the headhunting and slave-taking landscape of the Naga Hills. From this perspective *jhum* appears as another element in the trope of 'savagery', coterminous with more explicit forms of violence, a form of violence to the landscape and the natural environment: 'slash and burn' agriculture that left scars of '*jhum*med out' areas on the landscape. It was in this mode that the wastefulness of *jhum* was emphasized, portrayed as destructive, vagabond and wasteful. By contrast, rice production under irrigated management was valued as a superior crop of civilization – as demonstrated by other tribes in the region (Angamis and Apatanis in particular). These ideas resonate even in contemporary ethnographic writings on Nagaland and have entered public circulation through visual ethnography (Jacobs et al. 1990, 33–34; Strain and Ham 2003; Saul and Ham 2008: 157, 221).

As the colonial administration persevered the Nagas were further classified as 'wild' and 'noble savages', based on their adaptation to colonial rules and regimes of intervention. The 'noble savages' were the people who conceded to colonial administration and had ceased to take heads, slaves or engage in marauding raids. The Ao, Angamis, Lotha and some Sema areas that were made part of the administered Naga Hills became known as the 'noble savage' areas. The colonial administrators and civil servants admired the Angami way of cultivating rice and used it in moral judgements that distinguished the Angamis and other Tynemia tribes like the Thangkul and Zeliang from the rest of the population, who were dependent on *jhum* cultivation. John Butler, writing in the 1870s in his 'Rough Notes on the Angamis', described the terrace cultivation of the area as being carried to such perfection as to give 'a peculiarly civilized appearance to the country'. The use of a system of well-laid-out irrigation canals, the availability of organized labour in the cultivation of a variety of crops, reflected 'the remarkable skill displayed by the Nagas in their agricultural practices, a skill that would be received with surprise by those who had hitherto looked on them as occupying a very low position in the social scale' (quoted in Misra 1998: 3278, citing FPP1 841: FC No 106).

Describing the Angami wet-terrace rice cultivators, Major H. H. Godwin Austin, in his *Topographical Survey of India, 1873*, writes:

> In agriculture the Angami Nagas are far advanced. The principal crop is rice, grown on the hillside in terrace fields, skilfully and laboriously irrigated by artificial channels, often brought from great distance. It is not at all uncommon to see a hill thus cultivated from top to bottom, the whole presenting an unbroken succession of steps covered with a luxuriant crop. The description of land is much sought after, and is of great value. It is cultivated year after year, fertility being maintained by a liberal use of manure. The rice crop is exceedingly fine [and] nourishing, containing much sugar and gluten. The Naga rice owes its fineness to the natural richness to the decomposed clay shale, but they also manure at the breaking up of the soil in the nurseries, planting out just before the rains. A great deal of other cultivation is carried out upon the hill slopes depending upon the nature of rainfall. There is another system of agriculture called *jhumming*. It consists of simply a clearing of patch of jungle, and cultivating it for three years, after which the soil being exhausted, it is sufficient to relapse into jungle. *Jhum* cultivation is not much in favour with the Angamis, and is chiefly practised on the outer ranges; it is discouraged by the civil authorities as destructive to the fine forest of the hill district. (1873: 73)

This contrast is repeated throughout the colonial era. The settled peasant agriculture associated with highland wet-terrace rice farming is pictured as beneficial, while swidden farming and its associated nomadic mobility is presented as highly primitive. More specifically, sedentary highland wet-terrace rice farming is seen as indicating a higher degree of progress, sophistication and civilization. As against the taro, millet and Job's Tears grown in *jhum* farming, rice and its production under irrigated management, yet without the use of traction cattle, was valued as the superior, civilized crop.

Administrator anthropologists Hutton (1921a,b), Mills (1922, 1926), and later Fürer-Haimendorf (1938, 1969) in their various administrative and ethnographic texts applaud the benefits of wet-terrace rice cultivation over *jhum*. Fürer-Haimendorf goes on to show the contrast between wet-terrace rice and *jhum* through photographs published in the *Illustrated London News*[16] that had wider readership among the metropolitan public (Images 4.1 and 4.2). His representation of Angamis and Konyak Nagas through visual images adds to the idea of the ecological noble and wild or ignoble savage picked up by policymakers and post-independence anthropologists working in the Naga Hills. These seem to have rediscovered the Angamis as the only 'forest guardians', with their innovative *jhum* fallow management co-evolved with intensive terrace-rice cultivation, in contrast to the fallow *jhum* practised by the Naga tribes of present-day Tuensang and Mon districts of Nagaland, which involves burning of large swathes of forest.[17]

Private correspondence, another type of archival source from the earlier period, provides a striking moment. Mills writes in his 14 December 1936 letter to his daughter:

> We came down a long spur here to this huge powerful village, never before visited [...] They have brought us pigs and chicken, but their basket of rice had a thin layer of rice on the top and Job's Tears underneath, *a very inferior sort of grain.* (1995: 34, emphasis added)

In Mills's depiction, the coarser 'inferior' grain hidden underneath the superior but 'thin' layer of rice on top seems to be telling a whole story about a thin veneer of civilization (of course, this is how the Naga villagers chose to present their gift to him, and so may say more about their perceptions of civility and their relationship with the rulers).

In this administrative–ethnographic discourse, it is important that the 'superior' form of cultivation is itself indigenous – local – and not the result of external influences. Hence, the celebration of the Angami Nagas, located in the administered areas, for their terraced cultivation and associated sedentary way of life.

Colonial Tours in the Naga Hills

This section examines the context in which *jhum* was perceived, in terms of wider agricultural and revenue policies in the region. The argument being

Image 4.1 Irrigation marvels of the Angamis.
Source: *Illustrated London News*, 8 July 1937, p. 48.

developed here is that *jhum* policy in the Naga Hills was highly distinctive. It was different from the harsh and punishing swingeing policies adopted in other upland areas of British India. It was dictated less by colonial science than by the political exigencies of rule.

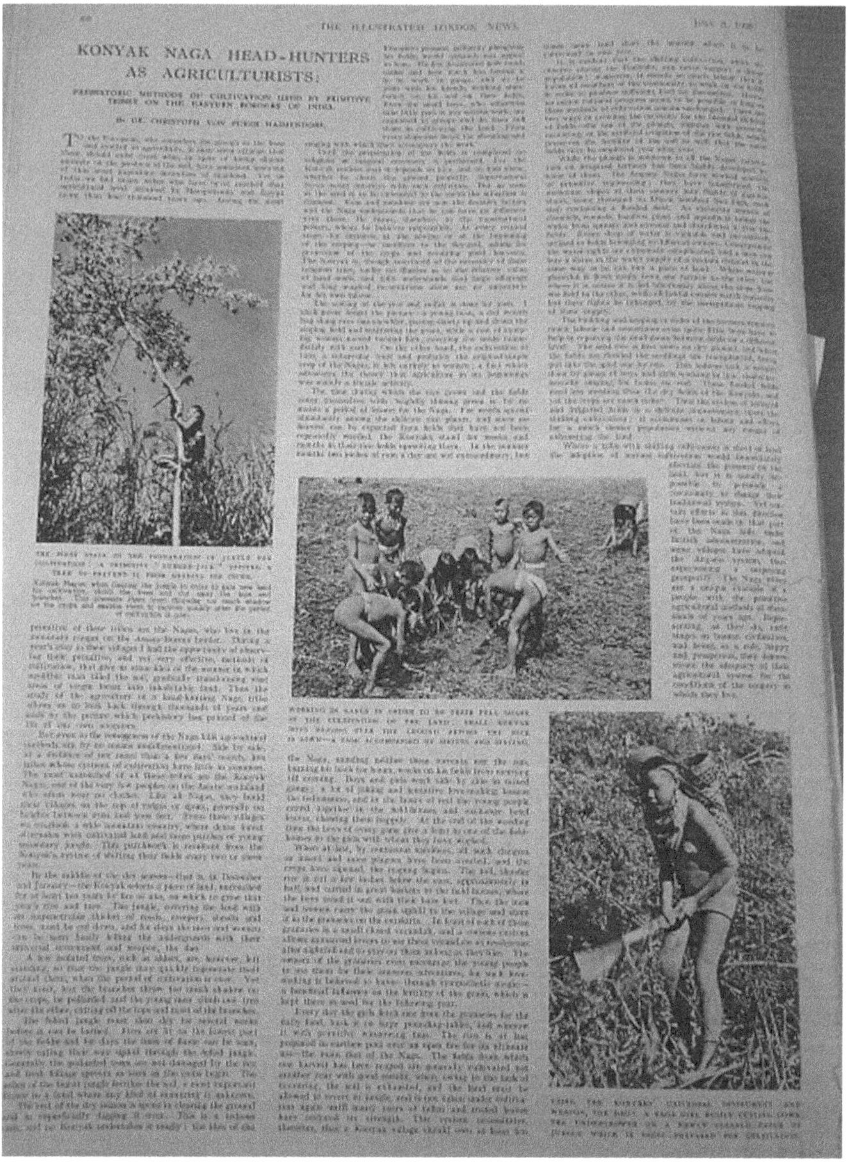
KONYAK NAGA HEAD-HUNTERS AS AGRICULTURISTS:

Image 4.2 Konyak Naga headhunters as swidden cultivators.
Source: *Illustrated London News*, 8 July 1937, p. 47.

Revenue-generation was never the goal of colonial administrators in the Naga Hills. It was never envisaged that the area might generate a surplus or even that it might be self-sustaining from the revenue point of view. Whatever projects for ‘improvement’ were undertaken, whatever tax regimes were instituted, these were primarily oriented to governmental aims. This is clearly stated, for example, in the 1869 *Annual Report on Naga Hills*:

> Government does not look at the Naga Hills as a source of revenue. The house tax paid by the Nagas and other tribes of the same class is to be considered in this light of an acknowledgement of fealty and token of submission.[18]

Paying tax was thus seen in terms of acceptance of authority and, more than that, a bonding of loyalty. Indeed, the annual reports on Naga Hills from 1869 to 1884 do not show substantial revenue collection. While house taxes were pursued more rigorously, they were marginal to the total revenue from the province.

Taxation was also a key mechanism in establishing indirect control through the embedding of village headmen for the colonial collection of revenue (house tax) from Naga villages. The *Annual Administrative Report of 1884–85* gives details of the house tax collected from the Naga-administered territories, which amounted to Rs 30,283. Of this, Rs 5,590 were paid as commissions to headmen of Angami, Kacha Naga, Sema, Lotha, Rengma and other miscellaneous *mouzas* in the Naga Hills.[19] The revenue assessments were Rs 2 and Rs 1 each household. The Angamis, some Sema and Rengmas were made to pay higher than the rest because they practised wet-rice cultivation, which was seen to be superior and more productive.

Similarly, although in the initial years the colonial administration depended on procuring cheap labour for the construction of pathways and facilitating the transportation of goods and services, the promotion of cash payment for labour services was seen as an instrument of submission and integration of the natives into market and trade activities based on currency exchange rather than a source of surplus. In the initial years, these inducements rarely attracted native Naga labourers, who were eager to return back to the wilderness. Money and trade had limited value for them, as social life in the hills was governed by barter exchange. Nonetheless, the introduction of cash very slowly but steadily transformed social relations, values and prestige and the nature of authority. This is reflected in the annual reports of the Naga Hills in the mid-1880s, which reflect the increasing integration of the administered Naga territories, their participation in colonial schemes of improvement and the effects of the introduction of cash payments for labour:

> The introduction of settled government in the Naga Hills has been followed by the development of trade, the cessation of inter village wars, an extension of cultivation, and marked improvement in the condition

> of the poorer classes. A taste for the luxuries of civilization has also been engendered, and Angami Nagas now invest in umbrellas, tobacco, and sugar. The great demand for labour has caused a large sum of money to pass into the possession of inhabitants of the district, and this has been invested in cattle and in the opening of new land, all tending to the increase in general prosperity.[20]

Saikia describes how, in the late nineteenth century, the Assam Revenue Department criticized the Forestry Department for overvaluing potential timber revenue in wasteland tracts, arguing that many reserve forest areas were of no value for forestry, in contrast to the huge revenue from forestry in the Brahmaputra valley. In the face of this criticism, the Forestry Department retreated, agreeing that 'in many of the forest reserves the value of timber was meagre and hence that they had been "unjustly reserved"' (2011: 274–75).

Despite the limited prospects of revenue generation, colonial administrators were interested in the promotion of commercial crops that would enhance trade activities in the hills – and thereby offer an attractive alternative to *jhum* cultivation. Colonial agrarian intervention in the Naga farming landscape started in the late 1860s, when they began to introduce potential cash crops, including tubers (primarily potatoes). The introduction by the imperial Agricultural Department of new crops from their experimental farm in Khasi Hills carried all over the frontier highlands from Khasi, Garo and Lushai hills to the Naga areas beyond the Sibsagar Inner Line in experimental farms and later in farmers' fields. The success of potato cultivation in the Khasi Hills prompted colonial officials in the Naga Hills to introduce it in the Kohima, Wokha and Mokokchung areas among the Angami, Lotha and Ao. Initial attempts at encouraging potato cultivation met with limited success. The reasons for failure were numerous and often lay in the delay of seed supplies arriving, as well as planting during the wrong season. During an experimental potato cultivation programme carried out in 1870 by Lieutenant John Butler, the officiating deputy commissioner of the Naga Hills, it was agreed that the introduction of potato cultivation in Koheemba, Jotsomah, Khonoma, Mozemah and Samoogoodting near Dimapur had proven to be successful. This prompted the deputy commissioner of the Naga Hills to buy fresh potato seeds from the Khasi Hills. In a letter sent on 17 February 1871, the Fort William authority sanctioned the dispersal of money from the deputy commissioner's budget grant at a cost not exceeding Rs 100 per mound (local unit of measurement, 40 kilograms made one mound).[21] However, subsequent plantations were constrained by mismanagement and failures on the part of the government in the dispersal of seeds. Reporting on the potato experimentation in the Naga Hills, Butler writes that the experimental plantation failed not because of adverse weather conditions but owing to the lateness of season when the seeds were planted out – despite which, he reported, the Nagas seemed not only willing but also anxious to give it a further trial.[22]

Cotton cultivation was also encouraged, although in this case it was not a 'foreign' crop, as cotton was already grown by the Nagas. Cotton cultivation was picked up very early as a potential item of trade. However, the quality, yield and variety were not very good. In the annual administrative report of 1869, the political agent of the Naga Hills writes that cotton had been found to be grown in the Naga Hills on the lower hills of Samogooditing. He had requested the Cotton Committee of the Agricultural and Horticultural Society to establish the quality of cotton available in the Naga Hills, indicating an early interest in consulting scientific committees on agriculture in order to ascertain productivity and yield. Cotton having been appropriated by the colonial local administration, seeds were supplied from government nurseries in Assam and Bengal for its widespread cultivation. However, like potato cultivation and other alternative farming practices promoted at various points during this period, these interventions were primarily carried out on the personal initiative of district officials in particular areas. When they were transferred more widely, the programmes faded away. *Jhum* also receives some particular attention in these late nineteenth century reports, indicating a degree of interest in scientific approaches.

Jhum Interventions

On 31 March 1881, G. A. Elliott wrote about the administration of the district after he had made a tour of the Naga Hills in the cold weather of 1880–81. He foresaw the great need of extending the practice of terracing as an alternative to *jhumming*, and he was very insistent that his office should try to spread this practice. Over 30 years were to elapse before these orders were carried out. Hutton began to introduce terraces into the Sema country about 1915. He also advocated the extension of the cultivation of potatoes (Reid 1942: 101).

The early twentieth century saw the colonial government taking a keen interest in native farming and adopting measures that could settle swidden farmers from their vagrant habits and semi-nomadic field-rotation practice. in the early decades of the twentieth century, as the colonial administration consolidated itself over the Naga landscape, officers in the Naga Hills engaged in many improvement schemes. However, there is little evidence of any systematic policy leading to the transformation of *jhum* landscape in the Naga Hills. This was because there was no permanent Agriculture Department in the Hill District and all activities were directed by the deputy commissioner and his subordinates, who were also involved in pacifying headhunting, resolving village disputes and settling petty crimes.

In his 1907 inspection note, R. R. Howell, deputy commissioner of the Naga Hills, highly recommended the pioneering work of Dundas, subdivisional officer at Mokokchung, in laying out terraced fields in the subdivision (Sema 1992: 100).[23] Angami Nagas, who were skilled in the development of terraces, began to be employed to impart their know-how to the tribes among whom terrace methods were newly introduced by the colonial administration. Thus, new fields were opened up with the help of trained agricultural demonstrators.[24] In 1910, Nathan, a subdivisional officer, with Mokokchung visited Longkhum village and, taking the *gaon burha* (village headman) along, selected a suitable spot for terrace cultivation (Sema 1992: 105). However, ignorance regarding the construction of a terrace and the variety of seeds most suitable for wet-rice cultivation tended to discourage the extension of terraces and so, an Angami Naga was employed to train the people (Sema 1992: 105).

N. L. Bor, the deputy conservator of forests, during his stay in the Sema country in 1920–23, had urged the Semas to conserve trees on their *jhums* and had encouraged experimenting with alder. Besides the promotion of alder-based plantation, the colonial government also introduced *lac* (shellac) cultivation. Among the Lothas, *lac* cultivation was promoted in the 1920s by M. Gregory with a group of people trained at Umran, in the Khasi Hills. In his report on *lac* cultivation in Naga Hills, he observes that the plantation seemed promising.[25] However, in spite of these initiatives *jhumming* remained the most important farming practice, only limited plantations and terrace cropping.[26]

In the 1920–21 annual administrative report for the Mokokchung subdivision, the subdivisional officer describes the progress of experimentation with terrace rice cultivation and the plantation of *lac* and fruit trees as measures of improvement in the Ao and Sema areas.[27] The subdivision office reports on these lines:

> Sitsoney Angami, demonstrator, supervised the experimental Terrace Rice Cultivation throughout the year. A sum of Rs 372 was spent on the pay of the demonstrators and his interpreter, and Rs 988 in making new terraces and improving old ones. Considerable further extensions were carried by Sema Sitzi in the upper Dikou valley and the Tizu valley [in the Sema area]. Similarly a sum of rupees Rs 527 was spent for the destruction of wild animals during the year.

The destruction of wild animals, including *mithun* (the wild Indian boar revered by the Nagas and used as a sacrificial animal), became central to the promotion of wet-rice cultivation, as the boar frequently raided the crops that were recently established in terraces. The tension between the farm and the forest was thus far apparent in the newly developed permanent terrace plots. If we look at the contemporary context, Nagaland has a very limited number of *mithuns* left, and their near-extinction is intrinsically linked to the expansion of wet-terrace rice cultivation areas in *jhum* fields. Applauding the work carried out in the Sema area, the subdivisional officer made detailed remarks, in the same annual report highlighted above, on the change of attitude towards terrace wet-rice farming of the native Sema *jhum* cultivators. The opening of new terraces was continued, and the area of operation extended. Writing on the progress of settled farming through terraces among Semas the report states:

> Semas are now beginning to realize that terrace cultivation gives a far more certain crop than any cultivation, and villages with two or three years swore that they would never abandon the ways of their forefathers and take to these new-fangled methods, are now opening up terraces. The yield from the terrace cultivation was good. The Sub Divisional Officer, Mokokchung reported that Bainho had an excellent crop from their terraces and were able to sell rice to their neighbours.[28]

During the preceding years, the Agriculture Department and the deputy commissioner's office had become heavily interested in promoting measures for conserving soil and forest through forestry and plantation work. The purpose of colonial administration was dual: to cultivate ideas of conservation as well as instigating the *jhummias* to settle agricultural economy linked to surplus production and sale.

During 1937–39 a sum of Rs 2,500 was spent in opening up terraces in Sema, Kacha Naga, Ao and Sangtham counties. In addition, Rs 917 was spent on pay and travel allowance for three temporary administrators.[29] This report was submitted in 1938 by G. D. Walker, Commissioner, Surma Valley, and the Hill Division by E. T. D. Lambert, deputy commissioner of the Naga Hills. In the report it was also mentioned that with the rise in population, *jhumming* continued to increase and if the forests are to be preserved with a view to preventing soil erosion and inundation in the plains, more funds would be necessary to induce people to take to wet cultivation.

Besides the efforts to extend wet-terrace cultivation, officials from the Forest Department contributed a great deal in promoting dry cultivation.[30] The Agricultural Department also proposed the opening up of a rice experimentation station in Kohima. The economist botanist from Assam, H. K. Nanda, also visited to inspect and aid the Agricultural Department that year.

He made valuable suggestions in his report for crop improvement in Kohima fields and the experimentation station. On the recommendation of the deputy commissioner of the Naga Hills, three agricultural demonstrators were sent to Upper Shillong farm to be trained in potato, sugar cane and fruit plantation. Accordingly, Visazo Angami of Viswema village, Entionen Jamir Ao and Panghangliba Ao were recruited for training in potato, sugar cane and fruits in Jorhat and Shillong horticultural stations.[31] By 1947, the Naga Hills District had one permanent agricultural inspector, two assistant agricultural inspectors, five permanent agricultural demonstrators, two temporary agricultural demonstrators and one office assistant to manage the department, intensify horticulture and terrace rice cropping among the *jhum*mias.[32]

Drawing on these initiatives, it became clear that the colonial policy to sedentarize *jhum* was now pursued with a more concentrated effort compared to the previous decades. This was driven by the Agricultural Department, which started building its strength in the Naga Hills and soon became more visible after World War II in the name of post-war rural reconstruction.

Colonial Policy on *Jhum*

By the late 1930s, colonial governments all over the empire were discussing the impact of shifting cultivation, and they had generated sufficient climatic data to produce a scientific discourse on the crisis posed by swidden (*jhum*) farming in the colonies (Rajan 1998). The links between forest fires, periodic flooding, *jhumming* and the threat it posed to settled farming on the plains were highlighted at the Hill Officers Conference held in Shillong in 1937, which produced a report subsequently published in the agricultural bulletin of Assam. While admitting that shifting cultivation was a cause contributing to floods, it was considered unthinkable that the hill people should be deprived of their land and livelihood in order to afforest the hills. A similar stand was taken later by colonial botanist J. R. Ainslie who, in discussing a resolution, observed that governments should 'take such measures as may be necessary to deal with the situation in cognition of the particular demand of each geographical and ecological area and political context of the empire' (Ainslie 1952: 386–87). In the Deputy Commissioners Conference, it was recognized that in some areas, such as parts of Naga Hills, North Cachar Hills and some parts of Manipur, there was reckless burning of land far beyond the land actually cultivated. *Jhum* practice was seen as irrational, and that there was the utmost need to check negligent burning. The conference resolved that steps should be taken to preserve a 'cap' of forest on the top of the hills and ranges and to induce *jhummias* to leave standing as many crops as possible in their *jhums* to assist quick generation while the land was lying fallow, lopping them when necessary in the interest of their crops.[33] Similarly, agricultural and forest officials took every care to make public that *jhum* cultivation should be stopped by writing pamphlets and short reports for public awareness on the effects of *jhum*.[34]

All these views culminated in 1946 with the adoption of Regulation III – the *Jhum* Land Regulation – which provided the 'land conservator' (broadly defined as the administrators in the districts and subdivisions) with a wide range of powers for the conservation of land and forest from wild fires and *jhum*. Violation of the regulation in any manner would result in imprisonment for either one month or a fine not exceeding five hundred rupees, or both. The impact of the regulation on *jhumming* could not be seen in colonial records. However, it can be said that wherever reserves were created, the ban on encroachment or interference with *jhum* was strictly prohibited.

In other parts of South Asia – Chotanagpur, the Chittagong Hill Tracts (CHT), Malabar, and the Karen territories of Burma – colonial conquest came much earlier than in the Naga Hills. The Naga Hills became full-fledged administered areas from the late 1870s, while in other places like Chotanagpur and Chittagong Hill Tracts, colonial control over the natives was firmly established during 1850–60. By the 1870s in Chotanagpur bullock cultivation was introduced on a systematic scale by the colonial government to contain swidden farming. Similarly, *tyangia* and *baiga chak* (improvements in swidden farming) were promoted by British Forest Department staff as alternative farming techniques and land use among the Karens of Upper Burma and the Baigas of Chotanagpur, Central India. In the Naga Hills, state intervention to contain swidden farming never picked up on a large scale. Periodic modification of the Inner Line and the subsequent redrawing of the Sibsagar District boundary with the Naga Hills excluded and encroached on virgin primary forests in the hills and protected them as forest reserves. In the colonial administrative phase, the major reserves were concentrated in Dimapur, Kohima and Mokokchung. Colonial administrative control was limited to the administered tracts. Mon and Tuensang were free areas contiguous to the Naga territories in the Burmese highlands (Saging Division and Kachin State). The independent Nagas were never administered. Swidden farming in the unadministered Naga areas was never disturbed nor truely influenced by colonial policies. These were the 'blank spaces', as Edmund Leach (1960) described the Assam–Burma frontier. Similarly, the promotion of settled wet-rice farming in non-Tynemia (Angami and Manipur Hill) areas through agricultural demonstration carried out by the Angami Nagas might have created its own contingency. Colonial records cite financial inconsistency as reasons behind its failure. As Deori writes, 'although the government was keen to promote the system of cultivation, it appears that the district administration had to work under considerable financial constraint' (Deori 2005: 48).

However, there can be other reasons behind such failure, which were negated in colonial discourses. My field interview with the villagers of

Tuensang reveals that in the late 1950s the Nagaland government sent some Angamis on a deputation from the Agriculture Department to teach the Tuensang people about changes in farming practices. When the Angamis returned from their training, the agricultural plots were abandoned, and many remained unused. The important observation was that creating terrace plots required a large labour force during the sowing and harvesting season, which the swidden farmers could not mobilize due to the constraints imposed by their own crop cycle in the *jhum* fields. Moreover, the productivity was low from these fields in the initial years of cultivation. The perception of productivity and labour varies from community to community in Naga society. The Angamis consider *jhum* as more labour-intensive, while the Konyak, Sangtham, Chang and Yimchunger Nagas consider terracing as more labour-intensive.[35] The different understanding of these two distinct farming practices raises new questions and challenges our analysis of the swidden economy. In Chapter 5 I discuss how these notions changed with a new land labour regime in the Yimchunger village. Similarly, labour relations based on kinship and *khel* (village ward) work parties were central to the land use and village economy (Fürer-Haimendorf 1969: 29–35) and remain to a great extent important to this day in *jhum* cultivation plots (as will be discussed in Chapter 5 on Naga land and labour relations). Labour could only be mobilized by chiefs, and the initial experiments were undertaken by village chiefs and headmen.

Similarly, promotion of trade in cash crops was a measure seen to bring about civilization among *jhummias*. In this context, a close reading of the Imperial Annual Administrative and Agricultural Reports published on the Naga Hills since the late 1860s will reveal that the administration was more interested in exploring the productivity of cotton cultivated in the hill districts. Several attempts were made to determine the quality, market price, variety and productivity of Naga cotton and their respective varieties found in Bengal and Arakan. The colonial administration was interested in cash crops and in the promotion of new food crops such as potatoes among the hill people. Thus, colonial conservancy efforts were mixed with ideas of improving farming through settled agriculture and the introduction of new crops and farming technologies that could produce a surplus for sale. In this way, the colonial government hoped to bring about improvement without confronting the local land-use regime.

The Grow More Food Programme of the Imperial Government

Just after the conclusion of World War II in 1945, there was a major push by colonial administrators on the Indo–Burma frontiers for rural reconstruction.

The Japanese war effort had destroyed British colonial patronage with the Naga chiefs of the frontier and had brought enormous hardship to the people. Even before the war began, leading frontier colonial administrators like Stevenson (1945) and Hutton (1945) had argued for the need to establish a better system of administering the hill people so that their loyalty could be sought in the interest of the British Empire in times of war and enemy invasion. These debates become more crucial during the early 1940s, when the boundary question between Burma and Assam came to be discussed by the hill administration.[36] This coincided with a drive for projects focused on increasing crop productivity and the promotion of horticulture, new crops and permanent settled farming through the intensification of irrigation facilities and the demonstration of wet-terrace rice cultivation techniques, including among the *jhum* cultivators of the Naga Hills. While this was in many ways a continuation of earlier modes of *jhum*-related intervention, there were also significant differences.

The key post-war intervention was the nationwide Grow More Food campaign. It drew its ideological impetus from the neo-Malthusian 'predicament of overpopulation' and was promoted in a number of countries by the Rockefeller and Ford foundations starting in the early 1940s. As Ross has provocatively argued (1998: 154–61), the neo-Malthusian thesis of a population-induced food security crisis, predicting apocalyptic outcomes of population growth outstripping agricultural production, was especially resonant in India where there was already a colonial legacy of presenting India as a Malthusian case. Ross analyses how this discourse functioned to draw attention away from the political, social and administrative arrangements of British land and revenue administration that were far more instrumental than population growth in determining the inability of Indian farmers to grow enough food crops to sustain the country's rising population that was leading to famines. Key elements of the problem included the intensification of commercial and cash crops, promoted to support colonial industries, and the demand for manufactured goods during the nineteenth and the first half of the twentieth centuries, together with the concentration of land holding in the control of the ruling elite. The Grow More Food campaign, at the same time, offered a palliative to agrarian unrest arising from marginalization and the underproductivity of Indian agriculture.

In the same vein, the Naga practice of *jhum* was branded an unsustainable and destructive form of agriculture. In 1946 – the same year as the *Jhum* Land Regulation – the Agricultural Department took the programme to the hill districts of Khasi, Jaintia and the Naga Hills. P. F. Adams, then the deputy commissioner of the Naga Hills, in his account for the *Assam Annual Agricultural Report, 1945–46*, observed that the government, under the Grow More Food

Table 4.2 Fund allocation to creating rice terraces in Naga villages

Name of village	**Amount**
Kilophutomi	Rs 140
Zumuthu	Rs 170
Khuivi	Rs 190
Sema area	Rs 655
Khuivi	Rs 190
Satakha	Rs 480
Sangtam area	Rs 190
Hoishe	Rs 120
Chishilime	Rs 200
Khukiya	Rs 115
Kilokhughoyomi	Rs 295
Sukhai	Rs 190
Tesangki	Rs 100
TOTAL	Rs 2,000

Source: Annual Report of the Department of Agriculture, Assam, 1946–47, p. 248.

programme, had appointed Mawsingh Kharsati as the agricultural inspector in charge of agricultural operations in the Naga Hills. Kharsati was assisted by five permanent and two temporary assistant agricultural inspectors and one agricultural demonstrator.[37] The programme during that period was one of the largest of its kind that the Naga Hills had ever witnessed. There were more areas opened under terrace-rice cultivation in the Ao, Angami and Sema areas. According to the deputy commissioner, 'throughout the district, including the Angami Area, the extension of terrace rice cultivation is going on rapidly because of irrigation projects taken up under the Grow More Food grant'.[38] A sum of Rs 4,000 was allotted for opening up of rice terraces in new areas. The amount spent is shown in Tables 4.2 and 4.3.

Similarly, a sum of Rs 15,000 was allotted for the construction of irrigation projects in the district. Water channels opened under the scheme made irrigation possible in areas hitherto having little or no water supply on account of streams being too far away (in many cases between two to four miles from the area to be irrigated). These channels enabled cultivators to open new terraces in areas hitherto unsuitable for this purpose. The amount spent is shown in Tables 4.4 and 4.5.

As a result of these projects, an area of approximately 9,000 acres was expected to be irrigated (with the possibility of expansion in future), and an output of about 1,500 tonnes of paddy production was expected. The deputy commissioner further observed that a great obstacle in opening up irrigation

Table 4.3 Mokokchung subdivision

Name of village	Amount
Ao area	Rs 655
Sema area	Rs 655
Lotha area	Rs 500
Sangtham area	Rs 190
TOTAL	Rs 2,000

Source: Annual Report of the Department of Agriculture, Assam, 1946–47, p. 248.

Table 4.4 Showing amount spent in Kohima Sadar

Name of village	Amount
Vishepu	Rs 1,000
Khekiya	Rs 520
Sukhai	Rs 1,010
Gukhwi	Rs 470
Khuvi	Rs 400
Ghokhuvi	Rs 200
Kilophutomi	Rs 350
Momi	Rs 50
Shoikhe	Rs 300
Kyakhu	Rs 150
Kivikhu	Rs 100
Hokiya	Rs 150
Thakia	Rs 180
Khughtomi	Rs 100
TOTAL	Rs 5,000

Source: Annual Report of the Department of Agriculture, Assam, 1946–47, p. 249.

Table 4.5 Showing amount spent in Mokokchung subdivision

Name of area	Amount
Ao area	Rs 4,926.14 paisa
Sema area	Rs 3,000.0 paisa
Lotha area	Rs 1,623.2 paisa
Sangtham area	Rs 450.0 paisa
TOTAL	Rs 10,000

Source: Annual Report of the Department of Agriculture, Assam, 1946–47, p. 249.

channels in the Naga Hills was the unavailability of blasting material to remove large boulders and to blast rock barriers to allow channels to pass through. A sum of Rs 2,500 was allotted for purchase of fertilizers and manure for free distribution, as Nagas had never used them in the past. As a result of the use of manure, extra yield was reported to have been obtained in Viswema and Khuzama villages. The report also gives details of trees and bamboo planted for anti-erosion work. New crops like sugar cane, oilseed, maize, soya bean, cashew nuts, sesame, *lac*, cotton, betel leaf (*paan*) and potato seeds were distributed in large numbers on an experimental basis. A demonstration garden was opened in Mokokchung subdivision in which interested children were given necessary instruction in vegetable gardening and raising fruit trees. The agricultural staff, while on tour, visited school gardens and gave advice to teachers and offered every possible help for improving school gardens. Also under the Grow More Food campaign, attention was paid to the rearing of livestock, as the hill people were semi-nomadic migratory cultivating groups. With the help of military dairy farms in Dimapur and Shillong, the Agriculture Department supplied a hundred head of Murrah buffalo. They were reported to be reared in the Angami villages of Ghaspani, Zuna, Sakhabama and Chachamia. Three Murrah breeding bulls were also procured.[39]

The overall impact of these agricultural extension programmes could never be assessed in real terms by the colonial government. In the post-Indian independence period, these ideas for replacing *jhum* farming were integrated into the five-year plans of the 1950s. The Grow More Food programme added a new threat to *jhum* cultivation: the peril of population pressure on the land. The emphasis in the post-independence period was on improving productivity. Thus, *jhum* came under the sceptical eye of the government on three grounds: first, as a primitive form of non-sedentary farming that produced more waste land than food grains; second, as a cause of heavy soil erosion and destruction to forest in the hills, leading to deluge in the plains; and, finally, as a threat to food security in the hills due to its unsustainable character. Thus, *jhum* came to be demonized as the single greatest evil in the hills.

The other feature of agricultural intervention in the hill districts of Assam and the Naga Hills in particular was cross-experimentation with local ideas of farming that were considered good practice. Such experiments brought many expected and unexpected results. In the Annual Agricultural Report of 1948, it is reported that the demonstration of paddy cultivation under the Grow More Food campaign had positive results in the Naga Hills compared to other hill districts of Assam. The report claimed that 'of all the Hill Districts, Naga Hills is the only area where irrigation and terrace rice cultivation (TRC) projects have great scope because of the cooperation and the responses of the cultivators to government help, and in fact Nagas have outstripped all other

Table 4.6 TRC Projects implemented in Assam Hills District and their relative success

Districts	Number of projects completed	Number of old projects repaired	Area to be benefited (in acres)	Increased production (in 000 monds)	Government's contribution
Khasi Hills	2	0	2	1.33	Rs 3,018
Jaintiya Hills	11	0	15.12	15	Rs 1,219
Naga Hills	178	10	10	780	Rs 2,055

Source: Annual Report of the Department of Agriculture, Assam, 1948, p. 245.

hill tribes in this direction'.[40] The statistics seen in Table 4.6 clarify the success of implementation of projects in the Hill Districts.

Nonetheless, even in the Naga Hills, success was coupled with failure. The report highlights the non-completion of eight projects and several other projects 'muddled' due to lack of grants and an inadequate labour force. However, these projects were geared to change the face of agriculture in the hills.[41]

In the shifting cultivation landscape of the Naga Hills, the colonial discourse continued to be expounded by post-independence Agriculture and Forest Department officials. This can be witnessed in the forest policy drafted soon after India's independence. While preparing the five-year plan chapter on forest policy, the Forest Secretary to the Government of Assam observes in 1954, writing to the Forest Secretary Government of India, citing Chapter XXI, Para 25, of the forest manual:

> Shifting cultivation is a problem confined to the hills, which are now under the control of the District Council. But the vital need of increasing areas of forest for their protective function in the Hills of the state has been from time to time as occasion offers explained to the District Council (*an institution set up by the 6th Schedule of the Indian Constitution for local administration in tribal areas of Northeast Indian states*) and tribal leaders. The question of setting apart land for permanent forest reserve in the hills is closely connected with the habit of shifting cultivation of the hill people for their maintenance and its solution lies in training up the people of the hills to permanent methods of cultivation and growing of cash crops to enable them to give up the practice of shifting cultivation (emphasis added).[42]

The statement clearly shows how post-independence Indian planners would see *jhum*. While the official policy on *jhum* has not changed, it has been made more forceful as *jhum* is depicted as 'redundant and anti-modern' with every *jhum*land legislation prescribing sedentary agriculture as the way forward to civilization.

During his stay in Sema country in 1923, Deputy Conservator of Forests N. L. Bor had urged the Semas to conserve trees on their *jhums* and to experiment with alder. Seven decades later, the NEPED programme eagerly looked upon alder-based *jhum* as one of the most sustainable alternatives to the indiscriminate burning of forest in the hills of Nagaland during its first phase of project implementation (1995–2001).[43] This NEPED department learned from the earlier practice of experimentation with alder-based pollarding advocated by colonial forest and district administration. Alder-based fallow management is generic to the Angami Nagas. The remaining *jhumming* tribes of Nagaland, while they have knowledge of alder trees, have not been able to develop such integrated *jhum* management plots as the Angamis. Similarly, in 2001, NEPED–IDRC external expert Malcolm Cairns carried out experimental and agronomic research to test the productivity of alder-based fallow management, and he subsequently wrote his PhD thesis in anthropology on this subject at the Australian National University. He observed that the fallow management practice was ecologically sustainable as well as more productive than the terraced farming carried out side-by-side in Khonoma village (Kohima).[44] His research findings were based on systematic records of climatic and soil data over a span of four years. He used advanced cartographic and digital devices like the Global Positioning System along with oral narratives and household field records to document and explain the rich heritage of Khonoma villagers and the experience of the Angamis in managing their *jhum* fields with systematic rotation and limited destruction of virgin forest. He and his PhD advisor have characterized this type of *jhum* innovation as another form of 'composite farming' (Cairns and Brookfield 2011: 58) following Rambo's (1998) use of the term in the context of the Tay ethnic minority in northwest Vietnam (1996, 2007).

Conclusion: The Debates on Swidden Farming

The chapter demonstrated how swidden cultivation came to be represented by colonial officials and the ways in which colonial frontier policies on *jhum*land management were determined by the 'science of empire' or administrative concerns. These days, policy planners use the 'destructiveness' of *jhum* and growing 'population pressure' on the land as rational ways to sedentarize *jhummias* in Nagaland. The roots of these concerns can be traced back to the colonial practices of *jhum*land management. Colonial policies could not change the very nature of farming, having only partial success here and there in a few village headmen's test plots and among tribes where colonial administration and the missionary influence was more pronounced.[45] Subsequently, despite the framing of new laws in the *Jhum* land Control Act 1946, state policies have not been the most influential force in bringing about agrarian change in Naga villages.

Nonetheless, farming today has changed remarkably in places close to towns, alongside national highways and in districts and subdivisions. The inflow of cash and growing opportunities provided by commercial and plantation farming has transformed the landscape. People no longer grow millet and Job's Tears, once the principal food crops of the region. Nowadays they are replaced by rice and other commercial crops, such as long beans, which were a minor crop in people's cuisine a decade ago.

What has brought about these changes if not solely *jhum*land legislation? As I discuss in subsequent chapters, drawing on my ethnographic fieldwork, villagers' changing moral values and attitudes to life have been instrumental in shaping new cultivation and land-use practices. Changing attitudes to work, time and labour are reinforced by new values attached to land and land relations reflected in the concentration of holdings by 'neo-tribal elites' (Rata 1999, 2000) in villages through privatization of proprietorship over communally controlled land resources. While tribal land in Nagaland is protected from alienation to non-tribals under Article 371 (A) of the Indian constitution, land alienation within Naga villages is becoming common as agroforestry and plantation programmes intensify to produce permanent tenure over *jhum* land. This becomes the topic of discussion in Chapters 5 and 6.

I also discuss post-independence electoral politics and the role of political party intermediaries and government representatives in villages who work as brokers in development programmes. Complementing this administrative and political penetration, the church has played a critical role in promoting new values attached to the consumption of rice as a crop of civilization (to be discussed in Chapter 6).

In Chapter 5, I begin by introducing the Yimchunger Nagas of the Shamator area, who became the focus of my village-based ethnography. I look at their social life and state interventions in agriculture that have influenced land relations and land use and have shaped the identity of these people.

5

Land and Land-Based Relations in a Yimchunger Naga Village: From Book View to Field View

> Traditionally, land and labor are not separated; labor forms part of life, land remains part of nature, life and nature form an articulate whole. Land is thus tied up with the organizations of kinship, neighborhood, craft, and creed – with tribe and temple, village, guild, and church. (Karl Polanyi 2001: 187)
>
> Landscape are cultures before they are nature; constructs of the imagination projected onto wood and water and rock [...] But once a certain idea of landscape, a myth, a vision, establishes itself in an actual place, it has a peculiar way of muddling categories, of making metaphors more real than their referents; of becoming, in fact, part of the scenery. (Simon Schama 1995: 61)

This chapter makes a transition away from what I have discussed until now on the history of Naga's relationship with the colonial state and the ways in which the frontier was managed through limited direct administration and tactical political intervention (annual pacification raids and building patronage through political go-between *dobashis* and village headmen). Here, I introduce the village where I conducted fieldwork and generated my field-based knowledge on *jhum* and the Yimchunger Nagas' everyday toil with farming, the land and labour relations, patronage with clans and the relationships villagers build with key village institutions – church, village council and the village development board. But more specifically, I shall discuss changing land relations that have become the basis of farming and livelihood for Yimchunger villages.

The village where I conducted fieldwork is located in the Saramati mountain range (Image 5.1) just 60–70 kilometres from the India–Burma international border and about 300 kilometres from Kohima by road. It was established in the mid-1940s by Shamator, Shiponger and Sanghpur villagers.

Image 5.1 Bird's-eye view of the study village.
Source: Author's photograph, 2009.

All three villages contributed land for the new settlement. In return, the first settlers who established Leangkonger had to offer *mithun* and a feast to the villages that contributed land. Villagers narrated that headhunting had continued in this part of Nagaland up to the 1950s. It slowly stopped with the preaching of Sema and Ao Baptist evangelical missions that toured the region since the early 1950s and instilled brotherhood through the message of the gospel. With the establishment of an Assam Rifles camp in 1951, and more recently the Nagaland Armed Police (NAP) camp in Shamator town headquarters, a new post-independence Indian state was permanently established. Headhunters were now suspected as 'gangsters', 'lawbreakers', 'rebels' and charged with sedition for challenging the legitimacy of the Indian civil and military establishment.

The landscape is dotted with patches of dense vegetation alongside *jhummed* plots, where one can see *jhum* rice, millet, corn, long beans – *kholar*, soya bean and other crops growing along the steep slopes of the village. The residences are located on a table top ridge where individual houses are built around the village road. The most prominent structure of the village is the Baptist Church, located in the upper *khel* (ward/sector), hereafter *khel*. The

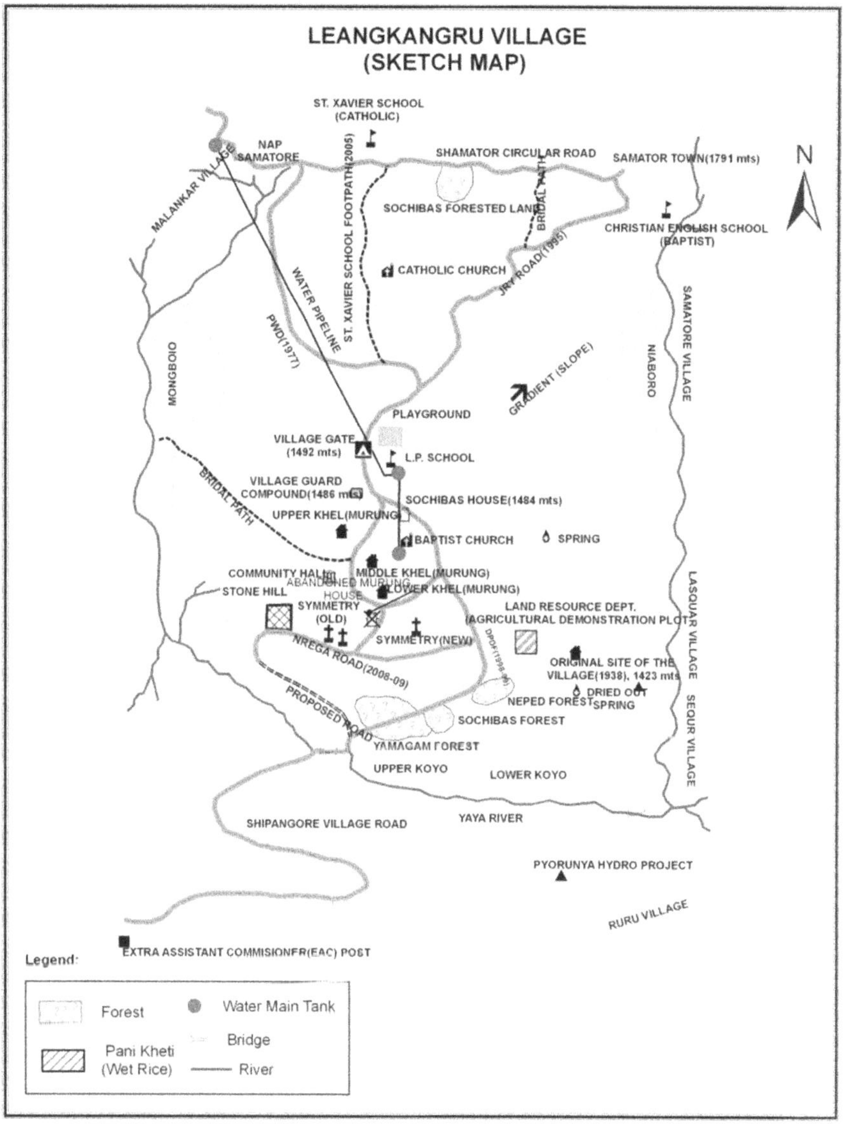

Map 5.1 Sketch map of the study village, prepared in the field with the help of village informants and host.
Source: Authors sketch map of Leangkunger village (prepared in the field with the help of village informants and host.

village is linked to the outside world by two main roads, one connecting the main settlements and the other circumventing the village and going downhill to join the neighbouring village, Shiponger. The village is divided into three *khels*, and these are represented by the three *murung* houses – *khiang yam*

Image 5.2 Murung house in the middle khel of the village.
Source: Author's photograph, 2009.

(male dormitories in the Yimchunger dialect). These days *murung* houses act as museums to showcase the memorabilia of the villagers, to memorialize their past and as a spectacle for visitors. In almost all villages, *murung* houses are decorated with moulds of human heads, *mithun* horns and log drums as cultural symbols of their past (see Image 5.2).

The village, as I soon discovered, had a fourth *khel* called *yamtüm* or 'corner *khel*' as it was located in one corner of the village. This settlement was established by a leading village patron who founded the new *khel* sometime in the 1970s and was successful in setting it up with his clan members from neighbouring villages and the NAP camp, who came slowly to join the new *khel*. All households in the *khel* are Catholic. Although both these segregated settlements are placed under one village council, there was little cooperation between the *khels* of the main village settlement and *yamtüm*. The self-proclaimed head of the *yamtüm khel* complained that they were facing discrimination in the distribution of development grants by the village council and wanted to form a new village. These allegations were reversed by the council office bearers, citing that the *yamtüm khel* settlers were mostly office-goers (government servants who earned a fixed salary unlike village farmers) and rarely participated in village welfare and agricultural activities. The reality of the fault line went much deeper, as these settlements were also divided by their religious confession. The village also has a government lower primary school and a playground at the entry of the upper *khel*, while the Catholic missionaries had established the English medium school and a church close to the *yamtüm khel*. The physical distance from the village to Shamator town was made easier to travel by the Public Works Department (PWD) footpath that was commissioned a few years ago to link the village with the town and the missionary school.

The villagers draw their lineage from five clans of the Yimchunger Nagas. Among them, *Jangrü* clan is the largest, followed by *Khiphurü*, *Jankhiungrü*, *Limkhiungkhiungrü* and *Kuzunkhiungrü* in the study village. The *Jangrü* clan or (*kuk*) as it is locally known in the Yimchunger dialect is further divided into four families who have their own village headman. The smaller clans *Khiphurü* and *Limkhiungkhiungrü* join the *Jangrü* clan headmen work parties during the cultivation season. The administrative geography of the village is shaped by three *khels* (upper, middle and lower) and is symbolically marked by three *murung* houses. Conversely, the social space is transcended by a household labour contribution that established clan- and lineage-based *khels* in the village.

The village also has a parallel institution called the 'Citizen', renamed during the village church silver jubilee celebration in 1999. Earlier, it was known as *Lenso* or 'New World'. The Citizen was established by Pastor Ranti Ba in the 1970s (Image 5.3). According to the Citizen president, it was an institution designed to bring moral improvement to the villagers, to improve work culture and to control households making beer. The Citizen pillar (memorial) inaugurated during the silver jubilee celebration in 1996 has inscription-names of all past village headmen and Citizen presidents. But, officially, the village council is the administrative and political decision-making body and has the sanction of the government. For the Baptist followers, the Citizen is the overarching village body that establishes the villager's morals. The village also has other groups like the Student Union, the Christian Youth Endeavour (CYE), church voluntary groups that play their roles in the village by the contribution of a tithe to the church and in generating good consciousness among the villagers through Sunday Bible teaching, music and other social events.

Agrarian Landscape

The agrarian landscape is a mix of *jhum*, newly established plantations, horticultural plots, long beans, relay crop plots and tree plantations. Similarly, the islands of plantation farms were disrupting the messy inter-planting under slash and burn. While in the lower altitude close to the local river popularly known to villagers as *yai* – irrigated rice fields – were cultivated by farmers,

Image 5.3 Citizens' pillar in the village.
Source: Author's photograph, 2008.

while terraces were built on the hill slopes to grow wet rice. The villagers practised both wet-terrace rice and wet-rice farming. The former was developed along hill slopes through government subsidies in the 1970s, as a rice-intensification programme was initiated through the appointment of *kelu babu* (agricultural demonstrators) who were sent on deputation from the Angami area to teach the Yimchunger villagers the art of wet-terrace rice farming. The latter was developed more recently in the later 1980s, opened up by farmers who received subsidies to protect their agricultural plots from the annual floods that inundate the low-lying fields along the banks of the river *yai* and were mostly wastelands (Image 5.4). The Irrigation Department subsidies in the late 1980s helped many villagers who became beneficiaries to open these plots and, today, with their dense irrigation networks, they have become the most precious land in the village, where the best rice is cultivated. In the last couple of years, people have also started using traction ploughs that are hired from the neighbouring Yimchunger village, Sekure, which has well-developed flatlands for paddy cultivation (Image 5.5).

Paddy field owners hold permanent rights to the land. There is no fallow period, unlike in the *jhum* fields, nor are there any communal obligations that compel farmers to give their labour to neighbouring households. The persons

Image 5.4 Wet irrigated field opened in the 1980s.
Source: Author's photograph, 2009.

who have the resources can pay wage labour and get their work done by hiring work parties from the village Citizen wage group, church voluntary groups, the Student Union, or even locally organized wage groups from individual *khels*. Marginal farmers who borrowed land from clan members are still dependent on household work parties and manual labour to grow rice in their paddy fields (Image 5.6).

The upland areas were *jhummed* with a bi-annual fallow cycle. Here crops of different varieties (Job's Tears, millet, maize, soyabean and *jhum* rice) are grown in mixed farms, inter-cropped, and are also grown as relay crops. Villagers explained that just before the 2008 church Golden Jubilee celebration, all the alcohol-brewing pots were destroyed. In the past, the church, Citizen body and the Student Union had fined villagers found brewing wine and stock-piling Job's Tears. Often villagers ridiculed households that kept Job's Tears and claimed that they were grown only with the intention of making beer. Rice is a mono-crop and is grown only in the first years (new fields), followed by millet in the second year in terraced plots. In fields where long beans are grown, they are relayed with corn. While corn is a soil-exhaustive crop, it is

Image 5.5 Traction plough.
Source: Author's photograph, 2009.

successfully relayed with leguminous long beans to restore fertility to the soil. The successful cultivation of long beans (*kholar*) in the village and its growing commercial value as a low commoditized crop (as farmers grow it both for household consumption and sale)[1] has made the village popular as a long-bean exporting village in Tuensang district.

In the late 1980s, government attention shifted from rice intensification through subsidies to the promotion of horticulture and tree plantation. This began with the introduction of pine, later followed by teak and *gmelina arborea* (locally known as *gamari*). Simultaneously, in the past decade, a variety of horticultural and cash crops have been introduced, from pineapples to oranges. Some of these have been successful, as farmers have started nurturing their orange and pineapple groves. Even the plantation of mangoes and bananas as fruit trees was carried out in the boundaries of paddy fields, often at the initiative of individual farmers who bought the saplings from Dimapur nurseries on their way back home after agricultural training programmes.

Land relations in the *jhum* plots were claimed by villagers as private. However, there were few communal features, unlike the paddy fields. Bi-annually, the village council demarcated cleared *jhum* plots, and individual

Image 5.6 Household labour in irrigation fields during the cropping season.
Source: Author's photograph, 2009.

households had no free will to start cultivation on their own. Households had to take prayers in the church and the *moharirrs* – a government staff person in the upper *khel* – was responsible for clearing the field, as I witnessed during my stay in the village. Secondly, the plots of *jhum* fields that were owned by individual households could not be divided between family members (sons) by parents. This was not the case in paddy fields, where a single plot was divided for sons. Thus, the relation of ownership was different between the *jhum* and terrace plots. This is also reflective of the tenure arrangement between *jhum* and paddy plots, because in the latter, no fallow periods are observed.

The communal nature of *jhum* plots did not make it entirely private property, although villages in their everyday talk claimed it to be privately owned and communally cultivated land. In the *jhummed* upland plots, both household work parties and wage groups organized by the church and the Citizen body took part in cultivation. The nature and role of work depended on what crop was cultivated, and also established a gendered division of labour. For the cultivation of long beans, farmers depended on wage groups, as this crop was more intensely grown and had a market value. By contrast, tubers and leafy vegetables were used as supplementary food crops and as fodder for pigs

Image 5.7 Man engaged in controlling irrigation water in paddy fields.
Source: Author's photograph, 2009.

(grown particularly by women work parties in the village). The kitchen gardens and mixed plots were mostly worked with household labour and primarily done by young married and elderly women. They had their organized household and *khel*-based work parties who visited neighbours' fields in rotation to plant tubers. In many of these plantation activities in which I participated while hiking up and down the village road to the agricultural farm, I could hear the rhythm of agriculture. Women sang while at work. Singing while hoeing the soil, according to my women interviewees, increased the urge for toil and made work parties efficient. Singing in groups while cultivating plots was interestingly absent in the paddy fields, where women went down to the leech-infested water compartments and engaged in backbreaking work, sowing the paddy in silence with a few young males around carefully controlling and checking the supply of water (see Image 5.7). The gendered division of labour was quite clear. While men were engaged in ploughing and levelling the paddy fields, women's work lay in collecting the rice saplings, tying them up and transplanting them. Men, on the other hand, were responsible for maintaining the water level in their paddy fields in order to ensure that crops were not starved of water.

Land Relations in Leangkonger: An Overview

Leangkonger village was founded in the 1940s by a group of surrounding villagers,[2] who thus became the 'first settlers'. They had to offer *mithun*, chicken and pigs, to Siphonger, Shamator and Sanghpur villages in order to establish the new village.[3] Having paid the *mithun* they claimed the right of access over the land resources and the right to be the owners of the newly founded village. The first settlers thus became the *atungrü* (custodians) of the village. I proceed with the question of land relations as it develops by focusing on the life story of one immigrant settler, Nikon, whose father's family was part of the first batch of immigrants who took refuge in the first settlers *khel*. The story of this new settler is important, as his family has become a very important link between the government and the village in the last two decades. Nikon's father migrated to Leangkonger in the mid-1940s from Sanghpur village, just after the raid carried out by the British punitive expedition of 1942 in which the head of his clan was killed. Nikon's grandfather had lost his life in the raid, as the family narrated. However, it was not the raid that compelled his father to migrate but clan infighting. Leangkonger had at that point just been settled. Land was not in short supply. In order to cultivate the agricultural plots, they needed labour – or, as Nikon's father explains, they needed labour and people to bring prosperity to the chief's farm.

Land relations in the initial period were skewed, with the new settlers depending on the first settlers for their subsistence. Nikon's father was given shelter in the lower *khel* by a headman from his own clan *Jangrü*. The headman gave him land for cultivation, in return for which Nikon's family gave free labour to the headman's farm. The clan members would collectively sow seeds in the headman's fields, an activity in which Nikon's family participated. Land was in those days communally owned.

In the olden days the village headmen were guided by a *khiungpuh*,[4] who had the customary right to first clear the *jhum* plots and observe *genna* (taboo) by not touching leftover food, after which people could begin cultivation. These *gennas* were important, as they were believed to bring fertility to the soil. The agricultural cycle revolved around the time reckoning set by the *khiungpuh*, who determined the sowing, harvesting and observation of rest days that were central in farm productivity. Labour was organized around communal household sharing. Families were grouped by *khel* under the leadership of the village headman, who distributed land to each family. In the *jhum* fields, the headmen received the maximum amount of farm labour, as they were the managers of agricultural production. The *khiungpuhs* received small tributes for making the animal sacrifice, with its powers to fertilize the crops, and for giving his blessing for a good harvest. Nikon's family earned a living from the single plot of land that was offered to them by the headman.

Among the new settlers in the village, Nikon's family was the first to accept the Baptist faith, as the *Jangrü* headman who gave shelter to his father was the first convert in the village and himself served as pastor for several years. Initially, the village had seven headmen and, over time, this number increased as more and more people settled in the village. In the olden days, land was not in short supply, although insufficient labour was a limiting factor for successful cropping. The chiefs were the most successful cultivators as they could mobilize a village household labour force to produce the surplus that was necessary to hold the large feast held annually after the harvest of millet and Job's Tears, which maintained their status among their kinsmen. These relations were soon to take a new turn.

Farming until this time had been restricted to millet and Job's Tears as staple food crops of the Yimchunger, along with vegetables and different varieties of tubers, which were cultivated in farmers' *jhum* plots and back gardens. Household food consumption of the farmers was not rice but was dependent on Job's Tears and millet.

The diffusion of rice consumption owed much to the missionaries who came from the neighbouring Ao and Sema areas in the 1950s. Rice was also promoted by regional council members in the late 1960s as they became nominated representatives, and by *dobashis*, who lived in towns and travelled to neighbouring district headquarters from which they diffused the new ideas of cultivation. The consumption of Job's Tears and millet slowly came to be ridiculed despite years of resistance. However, by the 1950s, the church had not firmly established its hold over the village. Then, after two decades of proselytizing mission work, in 1972 the village church was constructed. There were few pagans left in the village. Along with the new faith, new labour relations were successfully instilled in the hearts and minds of the villagers as well as the promotion of wet-rice farming.

During this period (starting in the 1960s), the government, through the Agricultural Department, also promoted schemes for rice intensification to create an alternative settled livelihood for farmers through the central Integrated Wasteland Development Programme. The rice intensification programme had begun under the North East Frontier Area administration soon after India's independence. It aimed to replace *jhum* cultivation with settled farming. The initial beneficiaries of the programme were the *dobashis*, village headmen and lower ranking government officials who were connected to the villages and acted as benefactors of these programmes. Other people, who were informants to the government, also benefited in the village.

The head *gauh burha* (village headman) of Shamator town had been one of the early beneficiaries of one such government programme in the 1960s. He had opened up terrace fields after being trained by the village-level workers sent by the Agricultural Department to provide instruction on the art of preparing terrace paddy field. These plots of lands were granted by the village headmen but were far away from the town and difficult to cultivate. In the absence of freely available labour, they were only taken up by farmers who could pay in cash or kind, for large farms required much labour during the sowing and harvesting seasons. Household work parties were not sufficient for such a collective operation.

The headman explained to me that, in the early 1960s, very few people like him took an interest in the cultivation of paddies. Labour reciprocity was based on cultivation of Job's Tears and millet, which were not only the preferred crops for household consumption but were used to make drink for feasts. Rice had little ceremonial use and formed a very small part of people's diet. The key obstacle to paddy farming was the labour 'shortage' – not in absolute terms but relative to the labour-intensive nature of wet-paddy cultivation.

It is only after the 1970s, when the village labour force was organized in 'age sets' by the village Citizen body, that rice cultivation started to pick up through the grouping of villages in terms of age groups under the guidance of the pastor. In the 1970s, during the construction of the church, the labour relations in the village were reorganized. The village was divided into age groups by the church to generate cash and a more productive labour force by engaging free man-days in public works programmes. The age group now engaged in earning wages for the church from a government-sponsored public works programme and by contributing labour during rice sowing and harvesting seasons. The intensive care that paddies require during the sowing season needed a dedicated labour supply, which was made possible by the age groups that gave fixed annual labour to the church by working in groups in paddy farms. The paddy fields proliferated and, with them, the right to permanent tenure was established in the lowlands, which had earlier been uncultivated and remained as thick jungles and swamps.

Government subsidies, a flood-control programme initiated by the engineering branch of the Nagaland State Agriculture Department since the 1970s and the development of minor irrigation channels helped farmers who could secure subsidies from the government to claim these lands from the village council and secure permanent tenure over them by growing rice season after season.

On the other hand, in the *jhum* fields the right to ownership remained communal until plantation farming started in the early 1980s. The role of the *khiungpuh,* who inherited the right to bring fertility to the soil, declined and was completely given up as the *khiungpuhship* was declared sinful by the church. As the whole village slowly adopted Christianity, the practice of offering millet and Job's Tears to the *khiungpuh* as gifts in return for his service completely stopped. Similarly, the 1970s also saw the proliferation of government-supplied guns for village defence. Cash incentives in the form of subsidies and people's demand for education and their desire for public employment brought in the circulation of currency through subsidies, loans, election-campaign funding and wages from employment in public works programmes.

Land relations were until then based on the village's first settlers being the custodians of land ownership. But the first owners had also offered land to the new settlers on the promise they would help in supplying labour to the village headman's house. Over time, the relationship between the first settlers and the newcomers started to shift. The change came with the coming of the church, the birth of a new village polity based on the government-recognized village council and later the Village Development Board constituted in the 1980s. These institutions were aimed at decentralizing development to the village bodies and have played a vital role in shaping values concerning land and property.

The evolving land relations in Leangkonger have to be understood, keeping in mind the changing social, political and economic transactions that have shaped clan and community relations. The clans that formed the first settlers in the village have the largest landholdings. Many late settlers who originally resided in the lower *khel* of the village with the first settlers (or landowners) have now moved to the upper *khel* (a *khel* that has expanded over the years). Influential second settlers among this *khel* have added private landholding by buying it from the landowning first settlers. The new settlers or latecomers in the village originally owned few plots of land, as part of the deal with the village headmen who agreed to give them *jhum* land for cultivation in return for labour. Household labour organized in village work parties were centred around *jhum* cultivation.

Over time, this relationship has been altered with the flow of cash and patronage built by new settlers with their fellow village members. Even today, labour relations have not been completely monetized through wages or any kind of payments in the village. Families and lineage members collectively go to perform agricultural work in the *jhum* fields to grow sweet potatoes, chillies or other kitchen garden crops. All crops are sowed and harvested with collective labour except rice and long bean (*kholar*), which require large numbers of paid hands. This paid labour is offered by church-collectivized wage groups, voluntary work parties and the Citizen wage group that involves all villagers. Without church-organized wage groups, rice farming had not been possible in the past. These new developments have created an ideal condition in which the beneficiaries of government-development programmes can invest in land for permanent development as well as buy new plots from distressed sellers.

A close analysis of current inter-household landholding reveals that the large concentrations of land are held by the first settlers who came from these three villages: Siphonger, Shamator and Sanghpur. According to the NEPED project survey, nearly one-third of the Leangkonger village farmers are landless (NEPED 1999: 109–12). While, according to my own household survey, nearly 30 families (nearly one-third of the survey households) reported landlessness and live as wage labourers in the village, participating as agricultural wage earners and doing non-farm activities. A small section of the population was engaged in basket making and in the manufacturing of machetes and guns (blacksmithing). In a few families, male heads worked as porters in the nearby Shamator subdivision (town) headquarters, while women worked on land leased from their kin or *khel* members and at times joined the female work parties organized by their *khels*. While in the past there was no sedentary farming practice, land was available for transfer to different claimants. However, once the islands of permanent forest and plantation farming began to take prominence, they obstructed others' claims to access or to have land transferred to them for cultivation. This development denied access to potential cultivators and has led to increasing pressure on farmland in the village.

There were others who consolidated landholdings in the village: town politicians and neo-tribal elites who worked as political patrons. Villagers showed me plots of forest and patches of timber gardens that were developed from land bought by the local member of the Nagaland State Legislative Assembly and townspeople who no longer lived in the village but acted as benefactors. The political patrons secured these lands through deals made with the village council members and their political representatives in the village. Often, clan affiliation helped to secure control over village land. I was able to document a case of land consolidation as narrated by the son of a local influential patron, who managed to plant valuable commercial trees in the area by securing land from neighbouring villagers. Over 17 years (1992–2009), he had planted 400,000 saplings over 370 hectares. Such large areas were planted in each successive season as land was bought from surrounding villagers who earlier practised *jhum*. Villagers in distress sold land to the local political patron, who provided them with much-needed cash and an incentive to grow maize and other crops for a few years until the tree trunks were large enough so that no food crop farming was possible.

How did this situation – or, rather, sequence of events relating to land use and land relations – come about? We also need to look at the longitudinal history of colonial intervention and the impact of plantation on land and land relations. This becomes the discussion of the next section.

Land Relations after the 1970s: The Rise of Second Settlers

Before the constitution of the village council, starting with the 1978 Act, authority was exercised exclusively by the first settlers. However, in the 1970s, elections were introduced in the Tuensang area and this gave second settlers or latecomers, who were numerically larger, a chance to challenge the leadership of the first settlers in the village. The first settlers' monopoly over diction making was broken by the village council, which was now represented by clan heads from second settlers and politically appointed village headmen, who drew their power and maintained control over resources by establishing links with government officials and political party patrons. The host first settlers were no longer in a position to maintain patronage relations with the second settlers, as 'customary' land-based patronage was fast declining, being replaced by government and development project networks that re-centred land relations.

Nikon's life story illustrates these changing land and power relations in Leangkonger village. As seen above, he belongs to a family of immigrants who were not the founding fathers of the village. Nikon was one among the few fortunate children in the village who was sent by a Korean Baptist benefactor in the early 1970s to receive a high school education in the distant Mokokchung district of Ao Nagas. By the completion of his schooling, he had become fluent in the Ao dialect, which, in his later life turned out to be a major advantage, allowing him to communicate with Ao Naga government officials and thus able to act as a go-between or broker in development programmes and other state planning activities undertaken in the village. He also came very close to the church. Nikon took up working his own field and established a family after the partition of his father's property among his brothers. His elder brother joined the state police as a constable and moved to Shamator town. The other brother joined the Village Guard with an honorarium of Rs 675 per month. Similarly, his youngest brother took the most prestigious position in the village, becoming the new headman of their lineage from the *jangrü* clan. He was appointed head of the clan members who had migrated from Sanghpur village in the late 1940s. It was a political appointment, as none of his forefathers were chiefs but immigrant second settlers. His sister was married to the Shamator town *dobashi* who later became the head *dobashi*. When Nikon's father's property was divided (which, according to custom, was done before the father died), each brother got one plot, or field. The eldest son, following the Yimchunger tradition, had the right to inherit the house.

On 20 January 1979, Nikon married Aiyasha, a woman from the village. It was one of the few holy marriages (held in the church) of the day. He inherited a plot of land from his father in 1978 and established his new home with his wife. On his return to the village from his high school study sponsored by the Korean Baptist missionaries, he became part of village institutions that were built around the church and the village council. In 1976, he served as the president of Christian Youth Endeavour (CYE), a Baptist youth voluntary group in the village; thereafter he served as its president for three years

1977–79 and from 1978, he held the post of Sunday School Bible Master in the church for two years. These positions were reinforced by his engagement with electoral politics and local politicians in the district.

He joined the Indian National Congress as village level grassroots worker in 1976. During 1980, he was selected as a member of the Village Development Board, a newly created village body with financial powers to implement development programmes and projects in Nagaland. All these positions brought him in contact with powerful men in the village and elevated his status among his clan members.

In the first year of his new family life, Nikon planted millet and Job's Tears as the main food crop along with maize, long beans and soya beans in a mixed plot. In the next cropping season, he repeated the same crops. In the 1980s he bought the first *pani kheti* (wet-terrace rice plot) by offering a lump sum to his neighbour. The farmer, as Nikon explains, sold the land to him in despair as he was in urgent need of money. The plot of land was first developed by a village headman through government subsidy, later left fallow and finally sold to its former owner in the 1970s. That year he cultivated wet-terrace rice. The terrace land development money had come from the grants he received as subsidies under several rural development programmes. Thus, he consolidated three fields into his landholding.

He served as a member of the Village Development Board from 1980 to 1986. As a board member, he was responsible for the mobilization of grants and the selection of beneficiaries. In the 1982 elections, the Congress Party came to power in the state legislature. Nikon and his clan members were actively engaged in development work as members from his native village (i.e., the village from where his family had originally migrated) and he became the elected legislator. The local Member of the Legislative Assembly (for the Shamator area) relied upon Nikon for his electoral success and his first victory after three successive losses in previous elections. Nikon became a strong figure in village politics. His position in the village slowly started changing from a second settler to a new patron, one who was educated and with strong links to the district administration, as well as a kinship relationship with the Member of the Legislative Assembly.

There were, in all, 29 families who had initially migrated to this village as second settlers from Sanghpur. They were mostly landless and had initially received *jhum* land for cultivation from the host village headmen. However, the relationship between the first and the second settlers changed over time as cash flowed into the village through development programmes, and educated villagers started participating in the government's development programmes. In 1985, Nikon, together with his brother (a Village Guard commander) bought a new wet-rice field. They paid a pig and Rs 500 to the villager, who was suffering financial distress. The cash was derived from the subsidies given by the Agricultural Department for paddy cultivation. Nikon also received subsidies from the Agricultural Department for levelling the field. The rice harvest that season was sufficient to sustain the whole family and very little millet was grown that year. The surplus millet he had stored from the previous cultivation season was sufficient to maintain the year's supply. His position as a Village Development Board member helped engage labour during the sowing and harvesting season on the rice paddy.

Then, in 1986, as the government began to promote agroforestry, Nikon first planted pine trees, which were supplied by the government's Forestry and Rural Development Department. Over ten years, starting in the 1990s the emphasis of the government shifted from promoting rice cultivation through subsidies and the diffusion of technology, including training and demonstrations of upland farm development through agroforestry. The primary emphasis of the government was on the afforestation and commercialization of local farming by introducing marketable plantation crops.

From 1988 to 1990, Nikon was a member of the village management board and chairman of the education committee. His rapport with the local management board won him many important positions that were normally reserved for village headmen. Nikon now found himself active in all development projects. He became the political henchman on whom the success of his party leader heavily depended. Nikon's brother was appointed as a village headman in the 1980s, and this gave considerable power to Nikon's family to influence village council decisions, as his clan and lineage in particular now were represented in the council. These positions brought legitimacy to their clan. Nikon emerged as a public leader as he acted as the main link between the village and various community-driven rural development programmes. He could command a larger labour force, which brought prosperity to his farm.

In 1987, he again bought a new plot of land on which he planted teak and *gamari* supplied by the Agriculture Department. The tree plantation programme that began in the early 1980s established permanent tenure over *jhum* plots, as trees take 15–20 years to mature. However, during my fieldwork, I realized that farmers did not mind destroying new plantations when alternative crop development programmes were offered by the government department. An irrigation survey carried out by the Flood and Irrigation Department officials, with whom I joined as a curious participant observer, revealed that villagers were ready to sacrifice the areas they had planted with valuable timber a few years back in order to replace it with irrigated wet-rice paddies. Village political patrons who had links with state officials saw new development projects as an opportunity to capitalize on the subsidies offered by agri-link projects, largely rice-cultivation and irrigation subsidies.

By the 1990s, the government's horticultural department had started emphasizing the improvement of upland farms through plantation agriculture, abandoning its sole emphasis on wet-terrace rice. The introduction of plantation crops, oranges in particular, and tree crops like teak, *gamari,* pine, *hollock* (terminalia myriocarpa) and *manjiam* (Australian teak) had significantly altered the tenure rights over the land. People who could permanently develop their plots were in no need to surrender their land to the village council. This was not possible for *jhum* land, because land that remained uncultivated in a particular season was transferred to another person who had the capacity to do so. In this way, a sense of proprietorship and permanent tenure emerged in plantation plots where the land became permanently tree-covered.

Nikon went on to buy many new plots in the years 1991, 1994, 1998 and 2003. The incentives and cash came from pre-election campaign money given by his party to win over votes and from funds that he received acting as a go-between in different development projects during his tenure as Development Board secretary. He purchased new fields individually from farmers who were in distress and in desperate need of cash. The council did not oversee such transactions but acted as arbitrators in times of conflict.

The promotion of plantation farming gave new opportunities to village patrons to build patronage links with landless farmers and those who owned just a single plot of *jhum* land. Households in my interviews reported that landed farmers took the saplings while landless farmers became the beneficiaries of subsidies by showing de facto ownership over others' land. This helped landowners like Nikon intensify their plantation efforts while subsidies granted to landless farmers met their emergency cash requirements. Another strategy adopted by the political intermediaries in the village to maintain patronage over the villagers was the distribution of tree saplings to their *khel* and clan members. This served two purposes: first to consolidate a patron's hold over the village household that belonged to his clan and, secondly, to establish patronage with *khel* members, which ensured a labour supply in the patron's field. Nikon directed the seed supply to his clan households that did not own land. He made them beneficiaries of loans and subsidies when he was the Development Board secretary and, in return, established his own plantation plots by securing free labour year after year, as the beneficiaries were obliged to offer free labour for being selected in development programmes.

The demand for cash came from farmers' desire to send their children to school. Our fieldwork data reveal that on average, a family in the village spends Rs 1,778 on children's education annually. Multiple landholding farmers send their children to private missionary schools in neighbouring towns and the state capital. This trend is then copied by marginal landholding farmers. The farmers believed it was only through education that they could escape their poverty. Nikon himself had sent all his children to missionary schools. With the financial support offered by the local Legislative Assembly member, he could send his daughter to study theology in Dimapur town. This help came just before the 2008 elections. Nikon told me that his family hoped that after passing her theology course, she would win a favoured place in one of the village churches as a women's leader. The other villagers see these moves as important steps in improving their livelihood.

By the late 1990s, Nikon's family had emerged as one of the largest producers of rice in the village, with the largest acreage under rice cultivation. Despite the failure of the monsoon in 1990, and a drought year, in the subsequent years rice had come to take a central part in Nikon's annual farming choice. During my fieldwork, he established ownership over 17 plots. Half his lands were under plantation and horticulture. Some were developed as test plots for biodiesel plants (*jatropha*) and he had only one plot under mixed *jhum* crops. He developed several orange orchards with the formation of Self Help Groups, which became the centre of government test plots, and which were registered based on his land as collateral. From here, 'pullies' or saplings were transplanted to other plots owned by his *khel* members. Farmers like Nikon acted as brokers to development projects. Land disputes that arose in the village were arbitrated by village headmen, who were aided by the village council members. The council chairman was often the most important actor in the village.

A large number of households had borrowed land from multiple landowners in return for a contribution of labour or cash payment to their fields during the cropping season. Normally, the owner would lend the land for two cropping seasons, after which it was his decision whether the tenant could continue. Tenants whom I interviewed said that these were inferior plots of land that were given to them for cultivation. The most fertile land was put to long beans and maize cultivation, which have become important cash crops. The relay cropping of maize and long beans had been successful in establishing the permanent tenure in *jhum* plots, where it was possible to grow long beans, as some farmers explained, for over ten years without any fallow period in between. Maize acted as livestock feed and as rations in poor households, where it was mixed with rice.

Conversely, land relations have changed over time with commercialization of agriculture and money transactions involved in the purchase and sale of land in Yimchunger villages. Both consolidation and fragmentation of landholdings was seen in the village. While Nikon and his brothers were engaged in consolidating land, others in the village were engaged in fragmenting their property. Zetero, the former schoolteacher and area council member, divided his paddy field among his three sons. This kind of division was not possible in *jhum* fields because of the restrictions set up by the *Jhum* Land Regulation 1944 and subsequent *Jhum* Land Act 1970 and more importantly Yimchunger customary law. For *jhum* fields, the authority lay in the hands of the person who had the widest control over the village. Traditionally, this was the *khiungpuh*, the first settler. However, these days it is the *moharrirs* (a government employer) who claims to have his influence over the largest *khel* in the village. He is responsible for announcing the clearing of forest just before the cultivation season. Although rights to property are private, land-clearing depended on some collective action. In the *jhum* fields, land use was based on customs and traditions that privileged the exercise of power by actors who had control over land use and labour. The existence of 'private' land ownership did not mean that individuals could cultivate their fields at will. This was only possible in the paddy fields that were opened in the late 1950s with support coming from the state. Here, the property relations were private, and the farmers had the right to decide on farming autonomously.

Meanwhile, the terraced fields were not covered by the *jhum* land laws and they were never cultivated in the past, being wastelands (annually flooded). These plots were recently developed as paddy fields, which required no fallow period as in the case of *jhum* plots. The individual investment on land for its development made them proprietors and developers of their land. The church has also played an important role in making rice popular in the farmers' fields. The increase in private landholdings, and individual ownership over land, primarily in paddies and plantations, has come about in the past five decades as the political structure has been shaped by new regimes of access and control over resources.

The access to communally owned land is controlled and determined by whoever can develop the land. The lowland paddy fields were, until very recently, wastelands in the village. But, today they are the most prized agricultural farms in the village. The village elders value upland *jhum* plots as 'wealth', but the changing cropping patterns and emphasis on wet-rice cultivation have made paddy fields more valuable and economically significant. The cultivation of paddies has not only sedentarized farming but has established permanent rights over land based on a farmer's capability to develop the plots. Plot development takes place through government subsidies and rice intensification schemes. Farmers who attained access to state schemes developed productive lowland plots. In recent years, rice intensification has become more exemplary, despite the state's shift in emphasis from lowland to highland farms. This has been brought about by farmers' increasing participation in lowland farming and their collective cooperation through dedicated labour. More intensive farming practices have taken shape both in crop selection and the nature of farming. During my fieldwork, Nikon invited people from a neighbouring village to run their traction plough in the village paddy fields, a practice that was begun only quite recently. The introduction of a traction plough was seen as an improvement over their pre-existing hoeing by manual labour, which produced inferior rice and disappointing yields.

The concentration of land has been influenced in recent decades by tenure evolution and the commercialization of agriculture. New commercial crops, and tree plantations in particular, have provided occasion for the concentration of ownership. Tree crops, because of their long rotation, produce especially interesting interactions with land tenure. The establishment of tree farms has been a critical step in the transition from shifting to sedentary cultivation, with trees defining permanent holdings. Tree planters take advantage of the almost invariable customary rule that land under crops cannot be reallocated, and staggered replanting makes that right virtually perpetual. Tree planting, as many studies have shown (e.g., Bruce and Noronha 1985), may initially be discouraged by insecurity of tenure, but once it emerges that trees will be protected, tree plantations can actually produce greater security of tenure and, more importantly, act as a means of claiming land. In this study, village tree plantation not only established permanent tenure rights but also denied others access to the plantations. They legitimized and sedentarized landholdings.

Nikon's life history shows how he became a landowner from a landless second settler by becoming a political patron and henchman in village development work. This was made possible by his participation in politics. From 1994 to 2000 he served as the Development Board secretary. During this period, he rarely stayed in the village, but was busy contacting project officials and Rural Development department staff for development projects that would benefit his farm through the creation of the Self Help Group and community development groups. In his leisure time discussions, he articulated that these Self Help Groups were meant to help each other develop. In practice, these decentralized and participatory programmes were helping patrons like Nikon to consolidate their holdings in the name of community, as each development scheme brought new assets into the village in the form of seeds, fertilizers, pesticides, equipment such as water tankers, sanitation equipment and housing materials. These assets went to the farmers who were responsible for monitoring and overseeing the distribution and implementation of the programmes. In the mid-1990s, Nikon also bought a rice mill. The mill came through an appropriation of funds from the Public Health Engineering Department's village water pipeline projects. He invested Rs 50,000 in buying the rice mill and a maize-grinding machine. These household mills have become a profitable business, as farmers often come to his house and de-husk grains for a fee. For many farmers like Nikon, the payoff from this off-farm income and contractual brokerage has provided cash income to increase their landholdings.

The church has also played an important role, not only in influencing consumption patterns towards rice but especially in contributing to labour organizations. In order to sustain paddy farming year after year, a large labour force is required. This is fulfilled by the church voluntary bodies, the Christian Youth Endeavour and other youth and fellowship clubs organized by the church. Nikon's role in the church was also a part of his rise as a second settler.

In 2009, Nikon had leased out two plots of land, one to a landless farmer from his own clan and the other to the former village pastor, his nephew. He did not ask for any cash in return; however, he expected farm labour on his own land from both families. The farmer looked after Nikon's paddy field and arranged for paid labour on this field during the weeding season, as he was away in town for medical treatment. The former pastor who borrowed one of his paddy fields was also a close kin member. His labour was used in building the paddy fields hut once the rice was successfully transplanted. These networks, built around labour transactions, were crucial to the success of farming.

Thus, with the coming of state administration and new legislation that formally recognized the Village Council, relations of power and reciprocity, as well as land relations were rearranged, meaning new landholding arrangements were produced. New categories of people (second settlers) rose to positions of influence within the existing social institutions in the village as they gained access to new forms of wealth. This also had an impact on land relations and is reflected in the landholding pattern in the study village. Before the advent of colonial administration, usufruct rights over land were stronger and established, as they defined the land relations between families. During my field study, these were shrinking, as more and more upland farms were brought under permanent cultivation through agroforestry programmes. The establishment of permanent rights over *jhum* land gradually destroyed the land entitlements formally established as a bundle of rights based on reciprocity between the chiefs and his subjects, the second settlers. The introduction of new crops like wet rice led to permanent rights being established over lowland fields, which once had been barren forestland.

Control over Access and Use of Land and Resources

In analysing the shifting land relations in Leangkonger – with the access of power transferring from the first settlers to the second settlers – I have drawn upon the inspiration of analyses in terms of control over access and use of land and resources. For African or in Southeast Asian highland societies, where land relations cannot be understood using the Western nomenclature of rights over land or tenure, a number of studies have defined the unique land relational and customary tradition of access to land (Berry 1988, 1989; Tsing 2002; Ribot and Peluso 2003). By contrast, in Northeast India, there is no study that engages with the critical debate on 'land', 'tenure' and 'rights' in the upland areas inhabited by 'Scheduled Tribes'. The understanding of land relations in Nagaland are based on Western notions of property and colonial classifications that were documented in colonial tour diaries and monographs.[5] Even – and especially – where legislation sought to recognize and protect 'customary' land rights, it was already modifying their nature, and paved the way for further changes. These standard classifications in the present land-tenure context misrepresent the complex power structure and labour relations exercised through land-based property relations. The egalitarian nomenclature used to understand land relations is clearly not applicable to the contemporary social context, where the village headmanship has been institutionalized in Nagaland. The customary land relations and principles of justice based on tradition are mediated by the district administration. The local *dobashi* courts do not operate within the mainstream judicial framework that compartmentalizes the judiciary and the executive. In fact, under the guise of 'customs', decisions are made by executives who interpret customs based on the knowledge of *dobashis*.[6] Faith-based institutions like the church also exercise enormous control over land-based social relations, as they control labour distribution through age sets that are organized in the village to provide free labour for these religious organizations. Similarly, land relations are increasingly being defined by the growing commercial value of the land and the growing trend of establishing rights to control and access over it.

In colonial Naga society, the collective right to land was seen as being embodied in the institution of chieftaincy. Thus, the Naga settler's right to the use of any part of *jhum* land depended on his political and *clan* allegiance to the village chiefs. He could not cultivate as an isolated individual even where he had the uncontested right of ownership of land he may have cleared. To imagine this relationship as 'merely' one of land tenure would be to reduce the nature of the relationship. It was a political affiliation that counted: the chief looked upon a man as a subject and not as a tenant, although the right to use land is actually conferred on the subject. This political relationship was in turn further politicized by the status the British gave to chiefs within their own political arrangements, including the power abrogated to themselves to appoint chiefs (formalized in the Chin Hills Regulations). In this way 'ownership' dissolved into administration: that is, the trusteeship, appointment and regulation of the users of land. Similarly, individual rights may now be considered as the way 'land' is used by individual members of the community.

The largely egalitarian social organization of Naga villages, where hierarchies were and are insignificant and provide a basis for sustainable resource use, helped construct the image of the Nagas as archetypes of authentic tribal social life. However, these images are not only at odds with the institution of headmanship but have changed over time with the value added to land and property relations.

Recent writings on land relations and property rights by scholars such as Sarah Berry (1975, 1988, 2004), Nancy Lee Peluso (1992), Jessy Ribot (1998), Charles Ribot (2003), Shabnam Gurani (1995) and Haripriya Rangan (1997) have outlined the very important aspect of land relations: that they are the socially constructed and historically embedded nature of property that allows rules and codes of property to be redefined and reinterpreted beyond the narrow legal confines. Berry's work on West Africa, particularly Cameroon, establishes that rights to land are embedded in multi-stranded social relationships that are open to redefinition under changing political, social and economic circumstances (Berry 1988). Land use and rural investment patterns can, in part, be viewed as a result of individual and group strategies to obtain and secure rights to land in this changing context. This emphasis is given by Berry to the centrality of power relations in the acquisition of land rights and in the process of social differentiation. Berry, in her study in Nigeria's cocoa belt, observes that rights to plant permanent crops were allocated separately from rights to plant annual crops, and were subjected to contradicting tendencies in terms of concentration and fragmentation (Berry, 1988: 61). Similar was the case of plantation farms in upland areas in my study village. Both long-bean plantations and tree-crop cultivation established permanent tenure over the land, which led to the establishment of permanent rights over it. However, fields that grew millet, Job's Tears, *jhum* rice and mixed crops were under a system of two-year rotational tenure. Farmers who could continue cultivation in the new *jhum* rotation got the land for cultivation; otherwise others could claim it.

These strategies helped many farmers consolidate landholding in the study area. Those who sold, lost or mortgaged their land became landless labourers, becoming in turn tenants and land cultivators. Labour has played a central part in regulating farming. In the new social setup in Naga villages, land concentration is closely attached to the social identity, power and prestige of the landholding patrons. Their social identity is achieved through maintenance of patron–client relationships with the landless and through brokerage in development programmes. Thus, land relations in Naga society have to be understood through a framework of multiple fields of power.

In the past, the village headmen were the most important landlords (but enjoyed only guardianship over it as custodians), who would make land available to other clan members during the cropping season, look after land disputes and act as trustees. However, in recent decades, land concentration has also taken place among new settlers who have built political patronage with landless households and their clan members, having limited plots of land for cultivation. The consolidation of landholding is also a reflection of land use. As the number of plots for plantation farming and agroforestry timber plantation and wet-terrace rice plots increases, permanent ownership, along with privatized land ownership, is becoming the rule. Forest plantations bring valuable income to farmers who sell firewood and often do not depend on the village resources, as they have moved out to towns or have family members as job seekers and public servants. While farmers who are landless or who have limited plots of land to sustain their livelihoods are constrained over the availability of firewood, as the right to collect firewood from private forests is restricted by the village plantation forestland developers. The circulation of cash through government poverty alleviation and agro-development projects brings in resources that aid in the building of new patron–client relations distinct from the ones shared between the chief and his clients. Thus, land relations have dramatically changed in the study village in the recent past.

Conclusion

This chapter has discussed the ways in which land and property relations have altered over time in the Yimchunger Naga village, with a shift towards private property. The colonial interpretation of land relations and the recognition of village headmen as village patrons gave the headman (chiefs) both legal and customary writ to make decisions on community land. With the upsurge of the money economy and state assistance for permanent settlement through the promotion of terrace cultivation, the value attached to land has changed, as it has become alienable: land has become a 'thing' that can easily be exchanged, mortgaged and permanently developed. The institutionalization of private property has led to increasing landlessness and to control over community land by those villagers who can establish that they have the resources to develop the land permanently by cultivating wet rice in irrigated terraces or by establishing tree plantations and horticultural programmes. In other words, it was via the promotion and adoption of terrace-rice cultivation and tree plantation that permanent individual rights were established and that the new regime of alienable proprietorship in which the value of land is determined through exchange and sale was put in place.

6

The Politics of Time: The Missionary Calendar, the Protestant Ethic and Labour Relations among the Eastern Nagas

> Last year, 1970, was a special year for the famous head-hunters in the Yimchunger area. Now they [have] surrendered before our lord Jesus, becoming his servant, abandoning old tasks. Once they were the instruments of Satan; now they are instruments of Jesus. (The Field Director of the Baptist Association, as quoted by Pathuvilian Thomas Philip 1976: 158)

> Believe in God. Don't drink rice beer, don't murder, and don't steal. (Words of Mr Ikavi Sems and Mr Nikoto Sema 20 March 1946. Kiussour Village Baptist *Boru* Golden Jubilee *Thiyukkian* 1952–2002: 12)

Both the above quotes that appear in the Baptist Missionary Records of Nagaland present a change that not only meant a change in local perceptions towards life, with the coming of a new religion, but a new ordering of time among the Yimchunger Nagas. In this chapter, I take the village church and its constitution of time as key to my analysis of temporal regulation of village life by the new faith. An analysis of time has now been used in numerous ethnological literatures to show its association with power relations (Rutz and Balkan 1992; Hoskins 1993; Bourdieu 1968). This chapter is about the politics of time: how, in rural-agrarian Nagaland, the Baptist Church shaped political and social relations through the regulation of time.[1] By centralizing and scheduling agricultural time, the church as an institution of faith emerged as the most important social entity in the village, regulating economic interactions and social transactions. I thus propose to show the dialectics of the time–labour relationship in Naga society.

It is now widely accepted in the social sciences that time is an organizer of power relations (Rutz 1992). Here I look at how the monopolization of agricultural time through the church schedules of calendric time regulates farming and allied farm-based activities (rituals and festivals) in a Naga village. Pre-Christian notions of time among the Naga tribes, particularly the Yimchunger, among whom I conducted my research, saw farming activities based on lunar positions that were decided by village calendric experts (who were the elder farmers themselves, members of the first settlers or founders of settlement). These farmers claimed to have tacit knowledge of seasons, the timings of sowing and harvest, by looking at the position and shape of the moon. Seasons were closely linked with the specific activity carried out during a particular month. Months were synonymous with the nature of work performed in the field. Days and weeks, in the past were calculated by using beads, by dividing cane sticks into pieces and by making knots in ropes made of leaves and fibres. Every Yimchunger elder whom I interviewed during my fieldwork described temporal record keeping, including the fixing of dates for meeting with other villagers, in this fashion.

With the diffusion of Christianity and the Baptist calendar in the late 1950s, the notion of time was reimagined, with the practice of observing Sunday as the rest day. The Baptist Mission calendar, based on the Gregorian system, also included particular days of fasting and celebration, such as Father's Day and Mother's Day, which were distinct from the farmer's local ways of connecting with nature (see Image 6.1, missionary calendar). The state also played an important role in popularizing the common Gregorian calendar and schedules for farming. Wet-terrace rice cultivation technology, introduced through the rice intensification programme in the early 1970s, brought in new notions of seasonality and techniques of sowing and harvesting that were not previously known. The new technologies were based on calendric dates of sowing and harvesting that were specific to particular varieties of wet-paddy cultivation. The farmers relied on government instructions and followed the calendric dates for sowing crops, as they were new to the landscape.

The major challenge before the missionaries was to invoke a new time regime and control over labour relations surrounding agriculture and crop production. This was done by establishing new institutions that were based on Yimchunger traditional social stratification of labour contribution within and between households, known as the 'age set'. The missionaries worked hard to instil new values, the 'Protestant ethic', a new work culture – 'economizing', 'thrift' and self-discipline by condemning feast giving. These moral values rearticulated the connection of time with practices of reciprocity, disinvesting the old forms of ritual power and authority.

In the contemporary context, younger-generation farmers no longer depend on the lunar positions to plant their crops. Similarly, the power to fertilize the field is no longer linked to the traditional social institution of

SHAMATOR TOWN BAPTIST CHURCH
2009 CALENDAR OF EVENTS

JANUARY 1st New Year 4th Fasting Prayer 18th Woman Harvest Day Programme	JANUARY Mo Tu We Th Fr Sa Su 01 02 03 04 05 06 07 08 09 10 11 12 13 14 15 16 17 18 19 20 21 22 23 24 25 26 27 28 29 30 31	JULY Mo Tu We Th Fr Sa Su 01 02 03 04 05 06 07 08 09 10 11 12 13 14 15 16 17 18 19 20 21 22 23 24 25 26 27 28 29 30 31	JULY. 5th NBCC Youth Day of Prayer. 11-12th Mission Week STBC. 18-19th YBYC Youth Retreat. 26th YBWC Shintbeshinbuh Münchi.
FEBRUARY 1st Handing & Taking 4th Wednesday Woman Day of Prayer 6-8th NBCC Annual KBA. 12th YBWC Executive 13th YBYC Executive. 17th YBBA Mission board. 24th SS Board. 25th Education board. 26th Central Executive Council.	FEBRUARY Mo Tu We Th Fr Sa Su 01 02 03 04 05 06 07 08 09 10 11 12 13 14 15 16 17 18 19 20 21 22 23 24 25 26 27 28	AUGUST Mo Tu We Th Fr Sa Su 01 02 03 04 05 06 07 08 09 10 11 12 13 14 15 16 17 18 19 20 21 22 23 24 25 26 27 28 29 30 31	AUGUST 9th Prayer of CASA CHS. 22-23rd Tankinberu Seminar Kewung Range.
MARCH 27th Feb-1st March Church Leaders Retreat. 3rd NBCC Fasting for Miglat Ministry. 7-8th YBYC Day of Prayer & Leader Exchange Ministry. 19-22nd NBCC Prayer	MARCH Mo Tu We Th Fr Sa Su 01 02 03 04 05 06 07 08 09 10 11 12 13 14 15 16 17 18 19 20 21 22 23 24 25 26 27 28 29 30 31	SEPEMBER Mo Tu We Th Fr Sa Su 01 02 03 04 05 06 07 08 09 10 11 12 13 14 15 16 17 18 19 20 21 22 23 24 25 26 27 28 29 30	SEPEMBER 3-6th Church Key Leaders Training. 13th Woman Thanksgiving. 16th Mong Belopi 20th Baptism & Lord Supper. 17-30th Woman Leaders Exposer Trip.
APRIL. 5th Palm Sunday. 6th-12th Passion Week. 7th NMM Board. 10th Good Friday. 12th Easter Sunday. 19th Woman Handicraft Day. 25th 26th YBYC Exchange Ministry.	APRIL Mo Tu We Th Fr Sa Su 01 02 03 04 05 06 07 08 09 10 11 12 13 14 15 16 17 18 19 20 21 22 23 24 25 26 27 28 29 30	OCTOBER Mo Tu We Th Fr Sa Su 01 02 03 04 05 06 07 08 09 10 11 12 13 14 15 16 17 18 19 20 21 22 23 24 25 26 27 28 29 30 31	OCTOBER 4th YBBA Mission Day of Prayer. 7th S.S & Child Board 8th YBYC Executive. 13th Church Advisor Board. 14th Mission Board. 15th Education & Literature Board. 22nd YBWC Executive. 25th YBWC Foundation Day.
MAY 1-7th NMM Mission Week. 10th Mothers Day. 10-17th Christian Home Week. 15-17th General Revival.	MAY Mo Tu We Th Fr Sa Su 01 02 03 04 05 06 07 08 09 10 11 12 13 14 15 16 17 18 19 20 21 22 23 24 25 26 27 28 29 30 31	NOVEMBER Mo Tu We Th Fr Sa Su 01 02 03 04 05 06 07 08 09 10 11 12 13 14 15 16 17 18 19 20 21 22 23 24 25 26 27 28 29 30	NOVEMBER 1st Sunday School Examination. 12-13th Church Audit. 13-15th Youth Annual Function. 20-22nd Central Council Sungphur.
JUNE 7th KTC Day of Prayer. 12-14th Children's Rally. 21st Father's Day. 23rd Pastor Pulpit Exchange.	JUNE Mo Tu We Th Fr Sa Su 01 02 03 04 05 06 07 08 09 10 11 12 13 14 15 16 17 18 19 20 21 22 23 24 25 26 27 28 29 30	DECEMBER Mo Tu We Th Fr Sa Su 01 02 03 04 05 06 07 08 09 10 11 12 13 14 15 16 17 18 19 20 21 22 23 24 25 26 27 28 29 30 31	DECEMBER 6th Advent Christmas. 7th Youth Annual Meeting. 8th Woman Annual Meeting. 9th Tenkihru Annual Meeting 14th General Meeting. 25th Christmas

Image 6.1 Yimchunger Baptist Church Calendar.

Yimchunger society, the *khiungpuh* family (the first settler who controlled the timing of the agricultural cycle). Rather, the preferred fixed dates of sowing follow the Baptist missionary and Agriculture Department calendar. It is only for sowing *jhum* paddy that highland farming is still dependent on lunar events. As one farmer recalled:

> If you plant on the full moon day, you will get good harvests but there is a high risk of pest infestation, while rice sown on the half moon day leads to an average yield but helps in protecting crops from infestation. A delay in planting leads to bad harvests.[2]

In the wet-terrace rice fields, the lunar calendar notions of half moon, full moon and quarter moon have been replaced by fixed calendric timing. Fixed dates are more important now, as farmers are becoming increasingly dependent on the organized labour force established by the church and plantation crops that come with government manuals and instructions provided by the Agriculture Department staff and demonstrators. The role of the church has become paramount; it has emerged over the years as the single redistributor of food grains in the village. The food grains that the church receives as tithes (one-tenth of the congregants' income in cash or kind) are redistributed in the village at a subsidized rate and bought by households. The institution of *khiungpuh*, the village headman (*gaon burha*) who had traditionally exercised authority over exchange by reserving the right to the first sowing of seed and receiving tributes in return – gifts as harvest from the villagers – is no longer upheld by the villagers (in fact, this practice is seen as sinful). Similarly, labour relations have been rearticulated with the formation of new associations in the village: church associations like the Christian Youth Endeavour – who pledge their labour as tithes to the church, during the Citizen's annual meetings – and associations such as the age group or age set (*thulan thulan*) that were central to every Naga society. As described in Chapter 5, the age groups or age sets are organized to pay free wages to the church through payment of 10 per cent as a church offering. Often these different groups work for wealthy farmers who live in Shamator town (5 kilometres away) and have the resources to pay wages to a large labour force.

In the study village, the Baptists go to the church and pray for a good harvest just before the sowing season. The same holds true when there is no rain. In remoter Yimchunger villages close to the Myanmar border, where Baptist Christianity has a much looser hold over people's lives, the *khiungpuh* still performs his traditional role in sowing the first seed. However, he is no longer entitled to take tributes from the villagers, nor to perform the rituals (*genna*) in the fields that are part of 'animism'. In fact, because the *khiungpuh* family are now Baptists, they have slowly entered the domain of church rituals and labour relations that define the terms of reciprocity based on the faith.

In the pre-Christian days, time was measured with the lunar calendar and crop harvest, festivals, feasts, headhunting rituals and crop-sowing ceremonies. However, now the church and its calendared schedule of a weekly rest day, fasting days and days of work act as powerful instruments of control and obedience, promoting discipline, punctuality and strict adherence to time now set by the village church. Even festivals that are linked to crop harvests and sowing are regulated by the calendric time set by the church. *Metümneo* (the soul-mating festival celebrated after the millet harvest in August) has emerged over the years as the most important harvest festival in the government calendar, although it is now rarely celebrated with community feasting in the village. *Metümneo* was traditionally followed by a thanksgiving in the form of free labour for the forefathers by all the villagers, devoted to clearing pathways that lead to the farmers' fields. Even this practice has been mediated by church calendric time. During my fieldwork, the thanksgiving ceremony fell on a Sunday, the day of rest, and so was postponed to the next day. Thanksgiving free labour organized by villagers has a social as well as practical significance in Yimchunger society, as it marks the opening of a new season for sowing the autumn crops (these days, long beans and soya beans are sown) after the millet has been harvested in August. It is linked to the agricultural rhythm in Yimchunger life.

The state, through its agency – the Agriculture Department – has diffused new crops, commercial as well as food, crops that require new understandings of seasonality, as they are not the traditional crops of the region and are unfamiliar to the farmers. Farmers in their turn have increasingly come to depend on calendric times and schedules circulated by the local officials in these 'backward' districts. Agricultural prosperity is now seen as closely attached to commercial farming. However, avenues for marketing such crops are limited given the geographical distance and practical problems linked to lack of transport facilities. Nonetheless, new crop combinations, such as long beans and maize and relay cropping of millet, along with soya beans have become popular, replacing the cultivation of Job's Tears and millet in the *jhum* fields. The long beans–maize–soya bean combination in my study area has transformed the very nature of circulatory movement under *jhum* cultivation ('migration' to the *jhum* fields) – another key temporal rhythm – by sedentarizing farming in many areas of *jhum* plots where long beans are cultivated.

While field rotation still continues, following a healthy ten-year *jhum* cycle, patches of *jhum* farmland have evolved where permanent cultivation has become the rule. The farmers continue to grow long beans year after year without burning new forestland. These have become permanent sites of annual long-bean cultivation. Farmers reported in my interviews that in their long-bean plantation, the system of plot–field rotation is not required as they have become permanent fertile grounds through relay cropping (a farming practice that allows crops to regenerate through nitrogen fixation by leguminous crops). For the new crop combination, the dates for sowing the seeds have changed and are purely based on calendars and schedules circulated by the state agricultural demonstrators and village-level government workers. However, when new fields are opened up by the farmers to expand their production base, the flexibility of sowing time is determined, rather, by the farmer's tacit knowledge and years of experience.

As Rutz (1992) argues, power is necessarily interested in controlling time but the relationship between power and time is contingent upon particular cultural and historical conditions. These contingent relations are now shaped by the church and the state institutions (Agricultural Department) that have a pervasive influence over native life. This chapter will explore these relations and how time can be imagined through the 'optics' of social relations or social time as an object of power.

Time Reckoning among the Yimchunger Nagas

Anthropologists have recognized that a relativistic view of time reckoning is essential to the understanding and description of non-Western cultures. However, this has also tempted anthropologists to place non-Western cultures, as Fabian has argued, as people of 'another time' (1983). I try to understand the ordering of time in relation to power by reference to the institutions of religion. Among the Yimchunger Nagas, the native understanding of time was in the past based on values that related to the day-to-day interface of villagers with nature, the agricultural cycle, the flux of day and night and the rhythm of life celebrated through feasts and festivals and explored through rituals and sacrifice. This is not meant to establish a romantic view of the past, but merely to establish that the local understanding of time was markedly distinct from calendric notions of time and labour introduced with the coming of the church in 1958.

With the acceptance of Christianity, there has been a slow collapse of many of the earlier practices. Today's farmers, for example, do not sacrifice chickens before the sowing season or offer 'feasts of merit' after a successful harvest. Rather, celebrations take place in individual households and in the church, following the Judeo–Christian calendar, with its three important Christian festival times: Christmas, New Year and Easter. People go to church and pray for a good harvest. Before the arrival of the missionary evangelists, time reckoning among the Yimchunger was dependent on the day-to-day needs of the present. There was limited planning for the future; agriculture and crop cultivation were designed to meet the population's bare subsistence needs year after year. In fact, farmers grew crops such as millet that required less labour, produced more grain and had a longer shelf-life. During my fieldwork, one of the villagers, who still possessed Job's Tears harvested ten years previously, explained that it was 'wealth' and a resource crop for subsistence during hard times such as famines.

Yimchunger time seldom indicates an exact point in the progression of time. Longer periods of past time were often explained to outsiders in terms of successive age sets and generations, for example, 'my father's generation', 'grandfather's generation' and so on. Beyond such expressions as 'yesterday', 'today', 'the day before yesterday', 'tomorrow', 'the day after tomorrow', 'nowadays', 'long ago' and the equivalent for month or year, other times in the past were indicated by referring to shared events, such as epidemics, droughts or a big forest fire (as in many African and Southeast Asian societies (Booth 1988: 81–91). The villagers narrated their past by referring to disease outbreaks (chickenpox, for example), as the reason for their relocating their settlement. It was very difficult for the elders to pinpoint the chronological order of events and how they migrated and settled over an area of land. Dates were not how the relation to the past was narrated. How, then, did the Yimchunger conceive of their future? Future events were rarely discussed in exact terms as to their temporal distance from the present. Just as with the past, it was common to reference points in time that were recurrent phenomena that affected their lives.

The lexical repertoire for the passage of time in terms of measurable amounts was equally scanty. A period of two or three days was expressed, depending upon the time scale, as a 'short duration' of time: this expression means a short duration as opposed to a long one, and not two or three units of time in exact terms. Temporal points in the past or future were rarely expressed in terms of measurable distances from the present, but by reference to socially shared experiences. This relational sense of time also applied to seemingly personal information. For example, during my fieldwork when I asked Joseph about his age, he went blank – but he could tell his relative age: that is, whether he was older or younger than someone else. This view of time, according to Newell, is humanistic (1975). What is important to the Yimchungers, for instance, is the rulers and what they did, not the number of years they lived as chiefs or village headmen. Thus, temporal duration is perceived qualitatively rather than quantitatively. These phenomena have also been recorded in other cultures in ethnographic studies on time reckoning (Ohnuki-Tierney 1973).

The calendric understanding of time and days was unknown to the people. People referred to the past in terms of generations. Leangkonger, the study village, was two generations old, according to the village residents, while the neighbouring village was five generations old. So, according to my respondents, the neighbouring village had a wider knowledge of Yimchunger origin and history. The knowledge of the past was central to their culture, customs and traditions, encoded in events not in chronology. Age sets were of primary value, far beyond their modern utilization for work distribution; rights and obligations within the group were crucial – but also for the concept of time as counted in generations.

The *Khiungpuh* and Control over Time

According to the Yimchungers' traditional narrative, there were three pioneers who opened new settlements during the inauguration of a village. These new pioneers, or village headmen, were locally known as the *Kiulingthusuphu* (village headman who carried the *mithun* while opening a village), the *Khihnuuhangbhu* (village headman who carried a dog while opening the field) and the *khiungpuh* (ritual priest who carried a hen while opening the field).

The *khiungpuh,* or village priest, possessed the power to fertilize the soil, using his ritual power to bring potency to the *jhum* fields. This authority was bound up with his role in the village as an owner, one of the first settlers and a reservoir of potency. It is through him that potency was imbibed in the crops. He unleashed and controlled the fountain of fertility power. The *khiungpuh* controlled time, which was integral to the cultivation cycle. It was he who had the first right to open the *jhum* fields for cultivation and had the definitive authority over agricultural timing.

The *khiungpuh* performed rituals and sacrifices with hens and pigs just before the onset of the sowing season and the clearing of a *jhum* field. For this, he enjoyed the right to receive tributes and gifts from his clan and village second settlers, who were obliged to offer tributes in return for his restoration of fertility to the field. The clearing of jungle was followed by rest days when no villagers were permitted to enter the *jhum* field. Anybody trespassing over these codes of conduct was punished by the spirits. The timing of the planting and sowing was based on his knowledge of the lunar cycle – he acted as a sort of soothsayer – and the synchronization of labour was done by the *khiungpuh.* As the custodian of the land, upon him was bestowed the customary right to redistribute resources for communal use. Thus, the *khiungpuh*, were the authoritative arbiters of agricultural time and timing. The institution of the *khiungpuh* was dominant in the village until the 1960s, when it came under attack by the missionary forces. (The decline of the *Khiungpuh* corresponds exactly with the growth of the Baptist faith that compelled church followers to abandon the rituals associated with sowing and harvesting presided over by *the khiungpuh.*)

The understanding of time in Yimchunger society is closely linked to seasonality and the cyclic events of nature. The rhythms are based on the work performed in a particular season. According to my Yimchunger informants, in traditional Yimchunger society, swiddeners followed the moon's phases to time their agricultural activities, dividing the year into 12 months of 30 days, as shown in Table 6.1.

The months are divided very specifically into 30 days according to the position of the moon, grouped into 15 'bright' days and 15 'dark' days. The farmers traditionally used the changing shape of the moon to time their sowing and harvests. Joseph explained to me how the Yimchunger forefathers followed the nature of the changing moon phases to time their sowing and cropping seasons. He drew sketches of various moon phases.

However, implicit in his description was an adjustment to calendric time. Although in the past the moon's phases were never calculated with fixed dates, his description was designed to match my understanding of calendars. The

Table 6.1 Yimchunger 'Forefather' Calendar

1	*Yungfu*	(January)	Resting month and drinking time, non-farming season.
2	*Khirim*	(February)	Clearing of fields, slash and burn.
3	*Khishang*	(March)	Preparation of fields, sowing of ginger, colocassia and other tubers. Tubers are an introduced crop in the study area associated with the arrival of Christianity and Ao missionaries. The staple food grains were Job's Tears and millet along with *alutho*, a local grain used for brewing.
4	*Aniu*	(April)	Spring: flowers bloom and nature comes to life.
5	*Khito*	(May)	New crops come up close to all the bridle paths to the field. Bridal paths are paths opened for communication and animal transport.
6	*Khihiam*	(June)	Transplantation of seeds and collection of new harvested tubers
7	*Langzu*	(July)	Busy collecting chillies; millet starts to ripen, along with corn.
8	*Muzhe*	(August)	New crops are ready, fields are cleared, millet and chillies are picked. The thanksgiving festival is performed.
9	*Muphu*	(September)	Maize is dried and carried in a *kiphu* basket
10	*Khipong*	(October)	The end of rainy season; people go to the forest to collect wood for constructing their houses (no longer practised as people have permanent houses).
11	*Hando*	(November)	Harvesting of paddies, soya beans, Job's Tears.
12	*Muliu*	(December)	Feasting and clearing of fields.

Source: Prepared by the author based on interviews with interlocutors.

first was the new moon, *khenu ahukhak* (10th–15th day), followed by the half moon *khenu atukhal* (15th–20th day), then followed by *amukhuk*, full moon (20th–25th day) and finally the *khenu ashaching*, last moon (25th–30th day). Joseph explained the relevance of the moon phases in farming. Maize, millet, rice and Job's Tears were planted in new moon days (10th–15th day); this was the phase favourable for the growth of the plants, and the harvest would be good. If planted during the half moon (15th–20th day), the plants are would not be good and the harvest would be bad. During full moon days (20th–25th day) no seeds were planted, as it would inevitably produce a low yield. However – as already described above – full moon is seen as the planting time when the crop is least threatened by pest infestation.

The *Khiungpuh* and the Synchronization of Festivals

The *khiungpuh* synchronized all agricultural activity right from the selection of field to the end of harvest time. In the traditional Yimchunger agricultural cycle, there were six important social events that corresponded to the sowing and harvest of crops. These social events were received by the whole village by holding a feast with the village first owners.

The year begins with the harvest festival *tsungkamnao*. It is a post-harvest leisure festival that was celebrated by various villages during the month of leisure and rest after a good harvest in January. The *khiungpuh* played a central role in the making of the harvest festival. It was also known as the 'friendship festival', as enmity between villagers was settled by feasting. These days, the festival is celebrated on 16 January by word of mouth.

The *tsungkamnao*, is followed by rituals that are performed by the *khiungpuh* just before the sowing season. The *khiungpuh* visits the field and performs a ritual sacrifice for fertility. The next day is observed by villagers as a rest day when no activity is performed except merrymaking and leisure. This is followed by collective village work in the field. The farmers grew millet and Job's Tears as their staple food crop with other tubers and occasionally rice. The millet harvest festival, also known as *metümneo* is also known as the 'soul-mating festival' among the Yimchungers. In pre-Christian days it was celebrated in the month of July. These days, the eighth of August is fixed as the celebration day. The fixing of dates for these festivities is linked to cultural programmes in state headquarters where politicians and government officials visit to inaugurate the festivities. The centralization of dates is important for urban Yimchungers, as month-long celebrations in different villages hinder work and everyday activities that are no longer attached directly to agriculture in towns and district headquarters. There are elaborate rituals associated with this festival's celebration that last for three days. However, these days, people visit the church and pray for their wellbeing and future generations. The *metümneo* festival is a great time reckoner. It marks the arrival of the new soul and offers thanksgiving to the dead souls. It was done in the past through symbolic remembrance. Five pieces of fresh meat were offered to a nature spirit. If the pieces remained unspoiled by wild animals, it meant the survival of the new born, meaning the survival of the new generation. If the pieces were harmed, the opposite outcome was signified.

The belief that the dead return tends to unite the past and the future. Just as yesterday and tomorrow have something in common in relationships today, so ancestors and descendants have something in common in relation to the present generation. The child may sometimes be referred as the father of his parent. The idea of reincarnation has been spoken of as indicating a 'cyclic' view of time (Elaide 1965: 89). Perhaps the Yimchunger concept is better described in terms of the coexistence of past, present and future. Here, I shall refer to Leach (1961), whose classic essay on time reckoning makes a clear distinction between Western 'linear' and indigenous 'non-linear' conceptions of time. Although Leach's paper provided a mechanical contrast between linear time and the festival markers of collective social life, his work has inspired other works, such as that of Victor Turner (1969) and, more recently, Block and Parry's *Death and the Regeneration of Life* (1982), which give very interesting perspectives on local ways of understanding time.

The *metümneo* celebration is a reflection of past and future events. The celebration not only marks the harvest of millet, but also ushers in the coming of a new generation. The *metümneo* festival is followed by *kulingaruiing*, the maize and *jhum* rice harvest festival. The farmers take harvest on the second week of October. The festival was celebrated to worship the weapons used in taking heads and in defending the settlements from enemy raids.

This is followed by the *rakrak pu neo* celebration, a festival that was supposed to determine child longevity. It was celebrated by sacrificing chickens. The feast was marked as an *anaogekhl* (peace-making) feast given between villages and an *akuntsukhhe* (prestige-gaining) feast within villages. These notions of time changed with the arrival of missionaries.

The Missionaries and Time Reckoning

With the early 1950s arrival of Baptist evangelists from the Sema area (Zunheboto and Aganato), the practice of time reckoning that was closely linked to the agricultural seasons and the age set changed for good. In the beginning, there was strong resistance to the change promoted by the missionaries. However, over time, the missionaries won the people's hearts and minds through their preaching, and as individuals began converting to the new religion. The Baptist missionaries brought in different values with a Judeo–Christian conception of time 'calendric time', and the disciplining of the native souls through the cultivation of faith, which villagers understood as the 'entry to the new world'.

The *khiungpuh* control over time slowly lost its charm over his people, as people started believing in divinity. The gradual conversion of the *khiungpuh* clan members to Baptism soon led to their isolation in the village. They became branded as hedonistic. The Christians' families were prohibited by the church from giving them tribute. This had a significant impact on their annual practice of fertilizing the fields through preharvest rituals. Villagers were encouraged to pray in the church rather than waste their resources pleasing the spirits. The *khiungpuh* thus lost more and more followers. The opening of paddy fields further limited the role of the *khiungpuh*, as wet-paddy cultivators did not require his blessings. In paddy fields, land was owned individually, and it was the responsibility of each farmer to open his paddy farm for cultivation.

The *khiungpuh* slowly became irrelevant in the village, like the 'brave man–headhunters' who took heads and enjoyed social prestige among the commoners. The recomposing of time symbolized a new set of values that distinguished the old from the new. The Baptist missionaries reinvented time by pitching their 'calendric time' against the local annual rituals of the native villagers that were based on day-to-day interactions and tacit knowledge of their agricultural cycles.

I narrate two stories from the study village to show how time was slowly centralized in the hands of the Baptist Church in the village, and how the native understanding of time was made less important in the lives of the people. This also marked the shift in control over time from the *khiungpuh* to the church. During fieldwork, people went to their fields, taking prayers and blessings from the village pastor during Sunday church service. The church pastor blessed the villagers for a good harvest. Potency now flowed from the Holy Bible. The scriptures show how, with the arrival of the Baptists, evangelist temporality was institutionalized by the pastors and their followers, displacing the institution of the *khiungpuh*, which became redundant as church time and the observance of the 'Lord's Day' took hold in people's lives.

Two Stories

The story I narrate was told by a Baptist Christian who was an eyewitness to these changes as he grew up in the village. The narrator is an old man in his mid-sixties. He introduced himself as the *khiungpuh's* brother. The man belongs to *jankhiungru khel* of Shiphunger village. I visited him with a special purpose to know about the *khiungpuh*.

Track 1

'Dear Abu, can you narrate to me the *khiungpuh* story?'

He smiles and starts.

> When a new settlement is established in a high mountain the *murung* house [male dormitory] is built first. Next the new settlers contribute labour toward the construction of the *khiungpuh's* settlement [village headman's hut]. The *khiungpuh* then places a live chicken tied to the murung house. The hen's longevity indicates the longevity of the *khiungpuh's* generation. The best hen is always picked and placed in the *murung* house. If the hen dies or is killed by stray animals, it indicates that the longevity of the clan will be short. On the other hand, if the hen survives it indicates the longevity of the clan.

As the house of the *khiungpuh* is built, the village is established. The *khiungpuh* has thus established a new village along with his co-inhabitants and clansmen.

The *khiungpuh* then visits the fields, selects a good spot based on the age of the vegetation and clears the field for *jhum* cultivation. He then builds a wooden platform with bamboo and a tree stump, where he sacrifices a pig. The blood is painted on the wood stump. Next, he shares the meat with his clan members just before sowing the seeds. The *khiungpuh's* clan gets the best part of the meat; the second settlers and latecomers share the leftovers. This is how first settlers maintained their relationship with the newcomers. The *khiungpuh* announces rest days just after the sowing rituals are performed. A similar rest day would follow just after the harvest of millet and Job's Tears. After a successful harvest is performed through collective work in the field, every household offers the *khiungpuh* a basket (*khang*) of millet or Job's Tears as a tribute. The tributes are generally a handful of grain given in small baskets to acknowledge his ritual power to infuse potency to the soil. Following a good harvest, the village chiefs then enter into competitive feasting to reconfirm their power and prestige among their kinsman. Often these feasts were also used as reciprocal measures to end feuds and to make peace with enemies.

In Shiphunger village the *khiungpuhs* were dominant until the 1960s. After that *khiungpuhship* went into decline. In February 1974, my interlocutor narrated, the converted Christians along with the church pastors went to the village *jhum* field in violation of the *khiungpuh's* orders. The *khiungpuh,* on hearing this news of the pastor's disobedience, chased the pastor and his followers out of the *jhum* field and confiscated their machetes and dibbing sticks. Subsequently, both parties were summoned to a meeting held by the village council. It was soon decided that within a year from the next cropping seasons, no *khiungpuhs* would visit the field first. The church by then had become significantly important in the Yimchunger way of life. By the late 1970s, as my informant narrates, the *khiungpuh's* family converted to Christianity, and *khiungpuhship* ended in the village. Today no *khiungpuh* are allowed to take feasts in the village. The church has become the centre of reciprocity and gift-giving through tithes (a one-tenth contribution of income from households).

Track 2

Abu next narrated a story about how the practice of building a *khiungpuh* house stopped in the village in the 1960s. Hearing that Hilimung, one of the khiungpuhs, wanted to revive the forefather's tradition of building the *khiungpuh* house, Hilimung accepted the labour of the villagers, but as a reciprocity and goodwill gesture. He did not feed the villagers. He abstained from giving the feast. The next day, Hilimung passed away. From that day people stopped the practice of building the *khiungpuh* house. They were no longer obliged to give free labour. In the following years, more people went over to the church, the government came and people devoted their labour to church work and engaged in the Public Works Department's road-cutting work. They devoted their toil to church work as jungle cleaners and as daily wage labourers.

Both these stories reflect important aspects of the Yimchunger past, the collectivization of work and the shift in ritual power and authority with the arrival of new values in society. These stories provide evidence of social time and how the control over time has a major influence on people's lives, the structure of authority and control over labour.

The arrival of missionaries marked a departure from the past, which was visible in the formation of a new institution developed by the church: the Citizen, an association of all villagers that had stepped into the 'new world', where 'economizing', abstinence from drinking, merrymaking and 'Feast of Merit' became the core values for improvement. These were based on ideas of non-violence and universal Christian brotherhood. The arrival of Christianity marked a break from the past. There was the construction of the 'other time'. The Judeo–Christian belief established a new chronology of events based on the Baptist missionary calendar, where time was condensed into a calendric schedule previously unknown to the Yimchunger Nagas. Initially, this created colossal conflict and a struggle for the control over time, as highlighted in the previous example in (Track 1), but in due course it established itself as an institution in all Yimchunger Naga villages and became a medium for difference between Christian and non-Christian.

The Baptist evangelist conquest of hearts and minds of the Yimchungers was different in practice from colonial pacification through headhunting raids. Yet they had two clear and equally similar purposes: to bring about civilization among the savages and to establish faith on a new system of time. Pathuvian Thomas Philip, writing on the spread of mission work carried by Sema and Ao evangelists, observes that 'Christianity penetrated in the Yimchunger area not by force, not by sword, but by the power of the word of God! Is it not wonderful that the head-hunters are becoming peace-loving Christians?' (1983: 158, emphasis added).

Philip further goes on to show how the headhunters were conquered. He narrates that the Yimchunger area was a den for headhunters. Thangpong was a notorious warrior famous for taking heads. He took some fifty heads for reasons of prestige. His escape in a scuffle with an enemy prompted him to join Christianity. Wangto of Wongong Village and Chanchi of Aiponger Village were also headhunters. Many of these headhunters from the tribe eventually accepted Christianity and gave up their headhunting practices.

Similarly, Khalenmew Yim, the author of *The Yimchunger and Christianity: A Brief History*, narrated the story of M. Yanthong from Leangkonger village, who was originally, like many other pagan Yimchungers, a heavy drinker. One night in 1958, a voice called him by name, and told him to accept Christianity. To this he replied, 'I drink two bamboo mugs of rice beer every morning, not to speak of how much I drink the whole day. How can I become Christian?' The following morning when he woke from sleep, he felt shy to touch or drink any beer from the mug. His father thought that he was physically unwell. His dislike for the rice beer grew day by day. Later, he accepted Christianity (2010: 27).

These stories were very popular and had become known in the area. Church followers would use these as examples in their 'exalting' Sunday Bible speeches to revive new spirit among followers.

Establishing Sunday as the 'Lord's Day' was one of the most significant sociopolitical moves made by the Christians in the village. It was a deliberate attempt to accentuate a social distinction that was based on the cultivation of new faith. The church calendar time was alien to the pre-Christian Yimchunger rhythm of seasons that followed a social time based on rituals and the cropping pattern of the community.

The major challenge before the missionaries was to centralize the organization of time and control over labour relations surrounding agriculture and crop production. This was done by establishing new institutions that were based on the traditional Yimchunger social stratification of labour contribution within and between households (the *khel* level): the institution of the age set. The missionaries worked hard at instilling new values, a 'Protestant ethic' – a new work culture – 'economism', 'thriftiness' to save wealth and self-discipline by condemning feast giving. The moral values were guided towards abstaining from wasteful spending and contributions of surplus to the church, in the form of tithes, that would bring about salvation and penance in their spiritual life and 'development' in the material world. Their energy was driven towards breaking away from a past that was traditional animist and pagan. The missionaries came from the neighbouring Ao and Sema areas and cultivated the notion of 'improved Christian humans' who presented themselves as advanced people, as they did not continue with their age-old practices. The consumption and promotion of rice was seen as a measure of improvement from wild crops, millet and Job's Tears, which were traditionally used in brewing homemade beer, now considered sinful.

The Baptist missionaries worked from the ground level. Their aim was to build the mission work from the grassroots. The Sema missionaries did the spadework in the Yimchunger area and were replaced by the Ao missions in the early 1960s as conflicts grew between the two associations. The Christian Fellowship in the Yimchunger village in the 1960s grew at a significant pace as they made contact with other Christian brothers across neighbouring regions, like the Garos. The Baptist associations were only strengthened in the late

1960s as the population slowly converted to Christianity. By the late 1970s, the Yimchunger mission had grown sufficiently large enough to establish its own association, separated from the Chang Nagas.

The Politics of Time: A Synoptic Illusion

The key to the politics of time is that it is a social and cultural construction: 'time belongs to the political economy of relations between individuals, classes and nations' (Fabian 1983: x). In this connection, Bourdieu's definition of understanding time through the calendar as 'synoptic illusion' (Bourdieu 1977: 97) is useful to understand the group activity 'collective action' as a codified aspect of social existence. The arrival of Sema and Ao missionaries demanded a new regulation of time over that which was already in practice. It meant an appropriation of the time of others, the institutionalization of a dominant time, and the legitimating of power by means of the control of time. Time, as Rotenberg argued, is shaped by power in two ways. First, in the labour process, planners who control the instruments of time control labour power. Second, the representation of time becomes the province of ideologies that legitimate the exercise of power (Rotenberg 1992: 33). The reorganization of time by the missionaries through the introduction of the Christian calendar cultivated a new dogma of faith for the organization of practices. The new Christian faith demanded that its followers discipline their livelihood towards 'the Lord's Day'. Sunday as rest day and a day for leisure and spirituality was initially hard to communicate. One of my hosts elaborated that the Christians would initially start their Sunday services on Saturday and completely miscalculate the timing. The old-generation church pastor explained that farmers were resistant to such idea, as they found no value in observing Sunday as a rest day. For them, the end of the sowing season and a successful harvest meant taking rest and enjoying beer with their kinsman by throwing a feast. Sunday made little sense for people who were cultivators and were engaged in games, headhunting, competitive feasting and blood feuds.

In the Ao areas in the 1860s, when the American Baptist missionaries first established themselves, people were visiting their fields on Sundays (the Lord's Day) to maintain their farms. During the same period in the late nineteenth-century Europe, E. P. Thompson writes, the Methodist Church was seen as an alien thing by the English working class. Workers were accustomed to using Sunday for a variety of feasts, fairs, labour recruitment and social visits. These were seen by the missionary workers as immoral and illustrated a lack of discipline in normal life. A century after this, the Ao and Sema evangelists were confronted with a similar mindset among the local people. Public morals in Yimchunger chieftaincies were encoded in reciprocal and obligatory kin relations between clans and service to chiefs in exchange for peace, protection and plenty. People lived in highland villages on the top of spurs, where they could keep a watch on enemy villages. Their livelihoods were sustained by migratory swidden farming and their social rhythm was consumption-oriented rather than driven by surplus production and accumulation.

All this was to change as the missionaries made their entry into the community and introduced a new time discipline. The Baptist time discipline became predominant within decades. During my fieldwork, I discovered a significant portion of the public attending the village church for Sunday worship. They had been disciplined to attend church service over the last five decades of evangelization. Very few people were uninterested in church activities, except those who were deviant and were often claimed by villagers as not being religious. Some of my young neighbours were reluctant to visit the church, as they felt lazy; instead they loved to spend the day roaming in the nearby town, returning back home in the evening. It was a day off from the drudgery of agricultural life. But the rest of the village stuck to the various prayer sessions, which also shaped my Sunday schedule of interviews, as I could not keep up with the number of services that were organized all day along, except for the first session in the morning.

In the pre-Christian days, the Yimchungers' world was inhabited by mountain gods, *gennas* and tiger-man spirits. There was no calendar or schedule and, hence, no calendar-based festivals. Yimchunger culture was one of much leisure. Among the most frequent activities, according to an elderly village man, were sleeping and eating, both of which were of long duration that were complimented by feast-giving celebrations after a good harvest. There were, in the missionaries' eyes, endless 'vacant hours' filled with tricks, dance, sports, conversation and sociability. A few decades after the missionaries' arrival, the diffusion of the Baptist time discipline in the Yimchunger land, and its integration into the power relations of chiefs and commoners alike was complete.

The Baptist calendar week became the time discipline. Days were given names in a weekly cycle, which revolved around 'Sunday' as the rest day and a day devoted to the atonement of sin, revival, blessing, worship and faith. Sermons were delivered at least twice a day. My fieldwork on Sundays was scheduled around the time set by the 'church bell', the eternal regulator of hours in a day. Sunday was special for all devout Baptists. The Catholics were a small minority, and they joined service in the yamtüm *khel* church that stood a kilometre away from the village. Sunday began with the morning prayer service followed by hymns, music, the chanting of the catechisms, preaching by the pastor and the Women's Leader and the Bible Home Master. The preceding days were busy with meetings, gatherings of youth, elderly, Women's Societies, Home Evangelist visits and Bible Study Groups. The afternoons and evenings were busy with prayer sessions and various group activities, such as singing practice and visiting people's houses to bless the sick and the elderly.

The basic institution of the Baptist week was the Sunday observance. The whole week, however, was suffused with routines of classes and meetings that became a part of the "Yimchunger way of life". School for learning scriptures was held at daybreak for one and a half hours. Women's congregations were held separately every Tuesday and Friday. Revival services became an annual week-long event to strengthen faith and the message of the gospel in people's hearts (see Image 6.2). The church pastor explained to me that revivals are the ritual of re-establishing faith. On Wednesday there was a church-leaders' meeting to assign the congregation's work for the following week.

In the 1950s, when the evangelists came to spread the message of the Gospel, the chiefs resisted missionary incursions. There were two waves of evangelic intervention in the Yimchunger area. The first missionaries to arrive in the Yimchunger area were the Semas, later followed by the Aos. As Yim writes, 'the Yimchunger heard the good news of Jesus Christ for the first time in 1936 (K. Ayiponfru village). The first church was established in 1947. Between 1947 and 1952, four churches were established with an approximate membership of about a hundred Christians' (2010: 22). The first Baptist Sema missionaries in Shamator subdivision established their evangelical work in a Yimchunger village called Sekur, which had a Sema Naga *khel*. The Sema evangelists found themselves secure among their Sema clan members and felt safe spreading the message of the Gospel in the territory of unadministered Nagas, who were still occasionally carrying out headhunting forays. In the later 1950s, as the mission activities grew with new converts, the church was established in Shamator town, the administrative headquarters of the Yimchunger people. One of the Khiamniungan Naga interlocutors narrated that headhunting actively stopped in the mid-1960s. Despite the headmen's displeasure at the first missionaries, some pioneers who accepted Christianity emerged in the late 1950s, and among them were the headmen's brothers. A decade after, the chiefs adopted the new religion, as they saw their kith and kin converting to the new faith, creating the unity of religion and power that has persisted to the present time. As one of the village headmen explained, the converts started organizing a village labour force and declined to offer labour to the chiefs, who were still pagan believers.

One of the villagers also narrated a painful story of when the missionary faith was first growing, and the chief's wife murdered her husband on the pretext that he was not giving up the drinking of beer and would cause obstacles to her participation in the church congregation. Many pious Christian believers who narrated the incident expressed that she had sacrificed 'the Satan'. An important difference to the earlier evangelical work carried by the American Baptist missionaries was the entry of indigenous Naga tribes into their lesser-known neighbouring tribes' territory. The baptized Ao Nagas, who

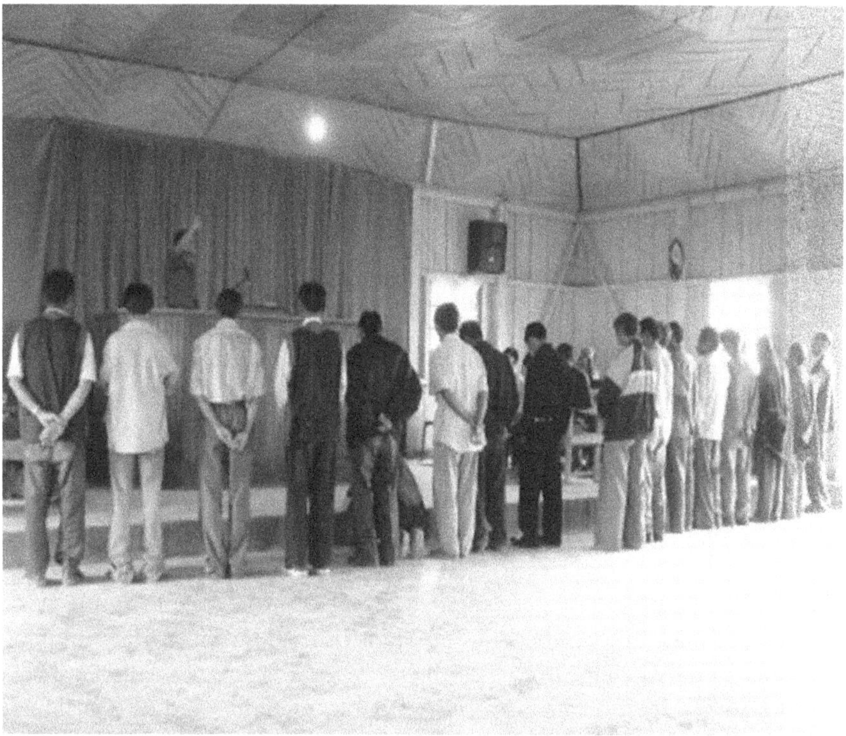

Image 6.2 Church revivals session 2009, Leangkonger.
Source: Author's photograph, 2009.

had a longer share of evangelical engagement with the Yimchungers since the early 1960s, not only preached the faith, but they also diffused new etiquettes, manners, behaviours, worldviews, and crops that were earlier unknown to the people.[3] For example, as discussed in an earlier chapter, the pastor Ranti Ba, who came to the village in 1974, established the institution Lenso ('New World'). The Lenso as a village citizen body codified rules and regulations for villagers, established the moral order and more significantly, regulated labour relations through age sets that were organized as age groups engaged in earning income for the church. These age groups contributed one-tenth of their earnings to the church. The village pastor during my fieldwork explained that Leangkonger was among the first villages in the Yimchunger area to adopt the institution of Lenso and had established itself as a model Christian village.

The church pastor, who came from a distant Yimchunger village, was very happy to share information about his lineage, stating that his mother was an Ao Naga woman. Both the Ao and Sema evangelists were seen as reformers and missionaries of goodwill and as members of a morally superior community who taught the Yimchunger the values of servitude and non-violence. They were responsible for ending headhunting and bringing peace. The new religion was equated with modernity, as an agent of improvement and the development of that which was old.

Similarly, the Huker Baptist Church Golden Jubilee *Khimtan Tsin* (history book), published in 1997, contains an insightful account of the influence by the Ao and Sema Mission on the Yimchungers:

> In Huker Village, Mr K. Kejingkhum, Mr P. Thsamphu and Mr H. Kiuthro had the privilege of attending school at Aizuto and Mokokchung in 1940. While they were studying in the Sema and Ao areas, along with their schooling they also grasped Christianity and a hygienic living standard. They received water Baptism from the hands of Mr Khuvikhe Sema. Thus, upon their return to the village, they taught and explained about Christianity and new habits they learned during their schooling.[4]

This process has also been experienced in other parts of South Asia. Hardiman's work among the Bhils of Madhya Pradesh shows not only how missionization took shape to convert the Bhils to Christianity, but also how certain Bhil forged their own relationship with missionary modernity (Hardiman 2008: 19–20). Hardiman observes from his other studies among the Bhils that 'assimilation to dominant values [...] provided a meeting point between the *adivasi* and certain progressive members of the dominant classes' (Hardiman 1981: 164). Similarly, among the Yimchungers, the Sema and Ao Evangelist associations were seen as agents of modernity to which converts related and through which started to imbibe Christian traditions. Every Sunday during fieldwork, I attended the village church Sunday Bible service. Village citizens, who were studying in towns pursuing higher education, were invited by the church pastor to deliver speeches to encourage fellow villagers to compete and excel in life and improve their existence with salaried white-collar jobs (government jobs). In the inspiring speech, the guest speakers would often refer to Aos, Angamis and Semas, citing them as rice eaters, as advanced Nagas, and urged church congregants to emulate these positive behaviours.

As the village headmen converted to Christianity, they became inextricably intertwined with church activities. The headmen surrendered their right to receive tributes and gifts that gave prestige and elevated them to a higher rank and established a hierarchy between 'owners' or first settlers, and commoners, second settlers. As they became Christians, the contributions that once reached the headmens' households were now directed to the church as tithes. The mission work blurred these boundaries. So, too, did the penetration of Indian state administration and the bureaucracy that made them a go-between in village development work. The mission work imbibed faith as the basis for reciprocity and prestige. People who contributed to the church were seen as just, pious and respected persons.

Prestige and power were now achieved through participating in church activities, and through acting as brokers in government-sponsored development programmes. The ability to gain a position within the church was central to establishing authority. The church was composed of dekons and kimzers who assisted the church pastor in his day-to-day activities. Dekons were selected from able-bodied men who were wise, just, morally upright and devout Christians. In the study village, when Ao pastors left due to some local crisis, which happened at many intervals in the past decades, one of the village chiefs renounced their village administrative duties to manage the church as a temporary pastor. In many villages, this became common, as pastors who came as the Lord's volunteers from Ao villages often had to retreat for personal reasons after a short voluntary service. In Leangkonger village, Yanthong, who was an ordinary villager, was blessed by the holy spirit and became a pastor with no formal training. As Yim writes, when his tenure as pastor in the village was over, many pastors came from Ao missions but could not stay in the village for long. So, in between, the responsibility came upon him and he served as a pastor for a record six times in the village (2010: 28–29).

Baptist Time Discipline on Sunday

By the 1960s the Baptist institution of observing Sunday as the rest day matched the administrative schedule of weeks, where Sunday was a holiday. Slowly the administrator's time came to match the church time. However, the agricultural cycle was still based on seasons. Sunday emerged as symbolic character of a day of atonement for sin and also as a day of celebration of the resurrection of Christ. A peculiar feeling that can be characterized as 'sombre joy' sets Sunday off emotionally from other days.

The tempo of Sunday is deliberate, a feeling that derives, in part, from the relatively long duration of a few activities. Sunday feels different because it draws a circle around the immediate members of a household or close kin. There is a withdrawal into the house away from public view. The intent of the day is to seek redemption through atonement. The day also marks a series of meetings when human labour is replaced by mental labour. Sunday observance is discipline of practice, proscribing some while prescribing others (Figure 6.1). Work activities other than minimal housekeeping, childcare, and some cooking are not condoned. Some people even visit wine stores in town and these activities are actively condemned. Time discipline centres on church

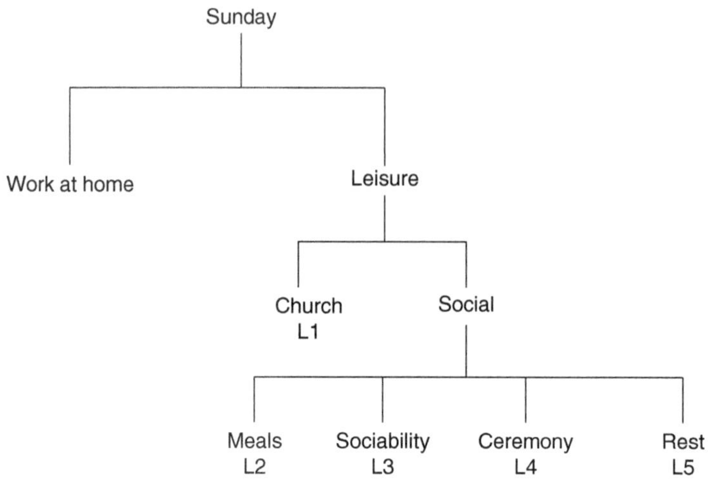

Figure 6.1 Index: L1 – Church, attending church service and meeting; L2 – Meals: Breakfast, Lunch, Dinner; L3 – Hanging around and visiting people's houses; L4 – Ceremony: Collective rituals related to church; L5 – Rest, lying down in one's own house during daylight hours, reading, listening to music.

attendance in the late morning and afternoon for adults and children. The duration of each service is normally an hour and a half in the morning and two hours in the afternoon.

The remaining part of the day is stipulated. After the morning church service, people go back to their houses. The Women's Leader with her group visits from house-to-house to restore faith. My host's wife was one of the prominent good Christians who visited people's houses with the Women's Leader. In the afternoon, after a short snack break, the village youth gathered in the church for service. There were cultural performances, church club presentations and announcements of important meetings. In the early evening before the last service of the day, the pastor announced a brief does and don'ts with the Bharat Scouts and Guide Cadets (mostly school children), who are trained to be disciplined for the annual Republic Day parade to be held in the town playground. The afternoon is separated from the evening by another church service, followed by an early evening meal before dark. Once again, the evening is different from a normal weekday: it consists of sociability with members of other households that can extend late into the night. The period of quietude continues as family members converse, listen to the radio, watch television, watch DVDs or perhaps read the Bible before retiring to bed. Beyond Sunday, all days are work days except Saturday evening, when prayer services are held.

In the early 1960s, the time discipline also spread through 'Print Baptism'. The Baptists took the utmost care to translate the Bible into the Yimchunger dialect; even preaching between villages varied due to dialectical variations. To cite one example, when the neighbouring Yimchunger village dominated by Tikhir-speaking people was invited to the village church in July 2009 during fieldwork, people found it difficult to follow the Tikhir dialect in the church prayers the village pastor said to introduce the Tikhir congregation to the church. Baptist missions have made their preaching intelligible to the Yimchungers by translating the Bible. The other denominations have also followed the same route these days. However, many of the aspects that were central to the village economy remained (labour relations) and could not be altered. The missionaries had to appropriate them in the context of a new world vision that was aimed at controlling time through labour relations. It also meant establishing control over the body and soul of every individual.

A New Order of Time: Land and Labour Relations

The new order of time was critical in monopolizing and restructuring labour relations that had previously been established through reciprocal inter-household and age set divisions of labour in Naga society. Communal reciprocal exchanges took place between the headmen and their followers, who lived in villages that were divided in *khels*. The missionaries were the first to bank on these networks of labour power. The construction of the Baptist village church in Leangkonger not only demanded a huge labour force but also regulated and regimented labour to provide free, dedicated and efficient labour to build the church. In the 1970s, the plan of the church building was conceived by the young and energetic church pastor, Ranty Ba. He was at that time in the prime of life. Married to an Ao woman from his native village, he was the most enthusiastic evangelist Leangkonger citizens had ever received from the Ao Nagas.

Middle-aged village men who remember him describe Ranti Ba as the most dedicated reformer among the Yimchunger. He instructed villagers to build houses with high roofs and proper ventilation. He was responsible for many other changes that the villagers fondly recollected. As a champion of missionary faith, he worked towards the modernization of the Yimchunger village. In 1972, the village church construction began and took almost two years before it could be completed. Skilled carpenters and mansion builders were Manipuris who came from the Imphal valley. The Nepali woodcutters who moved in with the British Army and later the Assam Rifles in NEFA days were primarily loggers who helped in felling the trees for the church's construction. Since the early 1970s, the Public Works Department was engaged in massive road building, primarily driveable black-topped roads that later became state highways. These roads were strategic to controlling the Yimchunger area when the military moved in after India's Independence in 1947.

Since the late 1950s, the Assam Rifles faced stiff resistance in Shamator from the Yimchunger warlords and NNC leaders, who had engaged the armed forces in guerrilla warfare. To dominate the area and to mobilize troops and food supplies, all-weather roads became a necessity by the early 1970s. The Public Works Department required people to cut roads. This event coincided with the construction of the church. It was during this time that the village work force was reorganized by the church pastor. In the 1960s, the government passed legislation on forced labour as a punishment to Naga villages that opposed the local administration and challenged the sovereignty of the Indian state. All these complemented people's resentment against the state while the community found solidarity in mission work.

Ranty Ba divided the village into age sets to attempt to teach the new cult of toil and hard work. He floated a new principle of organizing labour, called *thulan thulan* (labour organization based on age groups). The concept was already there, but the terms and conditions of how it operated changed. The *thulan thulan* were divided between age-sets (16–60) and were classified into six groups. They were put under a central governing body of an all-villagers Citizen – *Lenso*. It was led by a president, a secretary and a body of members all selected from the village, who were also members of the village council. The Citizen would soon become the all-pervasive and overarching institution of labour organization that decided on people's moral attitudes. It met annually, where rules, codes and ethics of conduct were framed for the whole village.

The Citizen acted as a moral authority in people's lives. It was responsible for all social activities in the village and issued birth certificates to children and new born, also registering deaths. It also had the authority to sanctify membership of the village clan. A couple of outsiders, mainly the two Nepali men whom I interviewed during fieldwork, became citizens of the village by joining the clans of their wives. Ram Bahadur took a new name after the initiation ceremony, 'Kimzer'. The Citizen gave him the new name as he became a member of the *Jangrü* clan. It evolved as an overarching body that covered the village council.[5] No individual was supreme to the Citizen. The *thulan thulan* were also part of the Citizen. The *thulan thulan* were organized principally to do efficient work and earn wages for themselves and the church. Each age group had to pay free man-days to the church in a year. Similarly, the *thulan thulan* was accountable for their conduct in public. They were obliged to pay tithes to the church from the earnings they made as daily wage labour in road construction and in the paddy fields. The middleman who engaged them in government contracts as wage earners also paid tribute to the church for the success of the work.

The church construction, which started in the early 1970s, took 24 months, an incredibly short time for such a huge structure to be erected in those days.

It was decorated with brilliant glass windows imported from the plains. During the Golden Jubilee celebration in 2008, a bell that had been custom-built in Kerala was ordered for the church. The villagers had contributed Rs 50,000 for the church bell through wage labour 'toil' and from household donations.

The year 1971 was also a drought year, and the village received aid from the government to overcome the food shortage. The villagers engaged in a wage-for-work programme, and subsidies were offered to villagers to open paddy fields. A lot of this public spending went to the church through organized labour and man-days given to the church to generate wages that were central for its construction. Since the completion of the church building, the institution of *thulan thulan* has been established in the village. To many of the Yimchunger elders, the denial that *thulan thulan* had ever existed in previous Yimchunger society reinscribed the symbolic capital attached to it. During my fieldwork I realized how *thulan thulan* and other organized youth labour groups in the village had transformed the agrarian landscape by timing agriculture with community labour or collective action.

Only farmers who could pay wages were in a position to employ these groups. They were mostly village chiefs, political intermediaries, party workers, government servants, and town dwellers who were engaged in business and contractual jobs but were connected to the village through their kin groups, who cultivated and maintained their fields with church labour drawn by mobilizing the *thulan thulan*. The remaining village farmers had to depend on the clan members and inter-household reciprocity maintained between families through kin labour. The inter-household bonds were weakening as more and more people were taking up plantations. In the *jhum* fields, such inter-household cooperation was concentrated on the planting of household garden crops, fruits, vegetables and tubers. In the farming of food grains, *kholar* – long beans and maize – wage labour was sought. In the paddy field, both types of labour operated. Small farmers reported that they used inter-household work parties, family labour and cooperation between families, while large amounts of labour were used by farmers who needed more hands in the sowing, weeding and harvesting season.

The Ao and Sema missions were a major influence in Yimchunger life as they learned from these neighbouring tribes the art of growing new tubers and wet-rice farming techniques. The pastors from the Ao and Sema community also brought along with them the culture of rice cultivation and spread the message of growing rice along with the teaching of the Bible. Inter-household labour that engaged in paddy cultivation was paid for labour. However, the relations of labour were not always measured in monetary terms. In most cases, except family and close kin relations, the labour was paid for in cash. This practice also strengthened in the paddy fields that were sublet and borrowed from the large paddy owners.

The church timing scheduled crop cultivation that could aid the production of rice. The church rice committee was not only responsible for keeping accounts of the tithes of rice received from the farmers, its redistribution and subsidization, but they also decided on the number of people who would contribute labour and on which dates. The church would not allow labour on Sundays. The regulated labour force placed the church at the centre of labour control. The labour, however, is shared between families, and labour is controlled by *khel* leaders; all other labour relations are controlled by the church. This labour contributes to the farm as well as government projects. During 2008–9, all farmers in the village were engaged in the MGNREGS government road-building project. In the study village, there were nine *khels* and each had its own household labour force. The *khel* leaders, who were also the village headmen, had to pay tributes of ten tithes to the church. The leaders were in return blessed by the church pastors during revival ceremonies. This reciprocal exchange of labour and faith is crucial for the generation of wealth and the functioning of the church. For example, the church constituted a reward for farmers who would contribute a maximum amount of rice as tithes. The families who could grow more rice implicitly engaged more labour from the church voluntary groups through contributing to church income. The households that were efficient in growing rice were recognized as hard workers and attained important positions within the church as dekons and kimzers.

The church did not break up the embedded structure of labour distribution in Yimchunger society but had to build on the existing layers of social relations by centralizing labour power based on ideas of efficiency and toil or 'hard work'. The division of labour based on age–sets was essentially towards maximizing efficiency, thus saving time and energy in the construction of the church building. Similarly, efficiency and time management are required during the sowing, pruning, levelling and harvesting of wet rice, as it is labour-intensive and rain-dependent. In the late 1980s, as *kholar* farming went commercial, farmers brought new lands in the village under their cultivation, which was devoted only to Job's Tears and millet farming. These proved to be the best long-bean cultivation sites as the yield and quality went higher. The *kholar* boom began in the 1990s. This would not have been possible without the new arrangements of disciplining labour that was organized by the church. *Kholar* was traditionally grown as a supplementary food crop in small plots of land. With the growing market demand, farmers started growing it as a relay crops with maize. The cultivation of long beans in large plots requires enormous labour power, as they are grown in *jhum* plots and cannot be substituted by work parties organized at the household level. Here, the church work parties have played a very important role in maximizing labour input during the sowing season. In recent years, farmers have also contributed land to the church work parties for the cultivation of *kholar* for market sale.

The church has regulated labour through the Christian Youth Endeavour, all unmarried women and men, the Village Student Union, the Citizen, which mobilizes the age sets, and the church's faith-based voluntary groups and clubs. There were two youth clubs organized by the church. The church emphasizes youth as the custodians of the future.

Time management in pre-Christian Yimchunger society was different in the ways people conceived of age sets as groups that had rights and obligations to the clan and lineage, as well as to the various *khels* they belonged to in the village. The village chief controlled the labour force through age sets as well as in his capacity to manage his men, the second settlers, and his close kin and family groups, who were obliged to him. The Baptist missionaries saw the age set as a resource for promoting ideas of improvement, bringing efficiency in work by grouping the village population into in age groups. This, as all my informants expressed, brought efficiency to the particular work that was done. The main idea was to use labour optimally. The age set division of labour was based on the principle of competitiveness. The villagers were no longer lazy and wary of time. As Rutz and Balkan write with reference to native Fijians, 'Fijian culture was one of "much leisure" (1992: 171). Among the most frequent activities were sleeping and eating, both of which were of long duration. There were, to missionary's eyes, endless "vacant hours" filled with sport, dance, tricks, conversation and sociability. When the missionary convert spoke of the High Chief Cajobay about enforcing Sunday Observance, the chief made a speech in which he described the situation of Fijian discipline thus: eat-sleep-bathe-eat-sleep-stroll' (1992: 171).

In the Yimchunger context, a similar condition prevailed. Labour was not seen as a tool or instrument of production but, as Janet Hoskins (1993) has shown in her study on Kodi time reckoning, it was for consumption, subsistence farmers were wary about achieving the surplus production they needed to secure their future. The Baptist time discipline was important initially to make followers realize the significance of the 'Lord's Day' and to discipline human activity around a calendar that had a chronological order of repeated events and was geared towards surplus production. The calendar thus became central to Yimchunger time reckoning. Farmers' cropping cycles changed as more time-oriented agricultural activities were popularized, like plantation farming, long-bean cultivation and, most important of them all, rice cultivation. Wet-paddy farming was a precision technology. In this practice, the management of time and labour are central to a good harvest. Toil and hard work was crucial for paddy cultivation and the post-sowing management of fields.

In this sense, the Baptist and government rice intensification programme were in harmony, as they both demanded the more efficient use of time and saw an improvement in the modernizing of agriculture through intensive and more efficient farming, based on the optimum use of land and labour, unlike *jhum*, where land 'went to waste'.

The Baptist order has played an important role in diffusing the idea of rice as 'civilization's crop'. Elsewhere, villagers claim in my interviews that rice has become popular due to its better taste than millet and Job's Tears. Wet-rice cultivation, though commercially less remunerative, has become the most important means of producing the staple food grain. It has the highest status value in the village. People who give more rice gain high prestige within the village church. Rice tops the crops that are redistributed by the church to the village households after the tithes are obtained from individual households after each harvesting season. Villagers who do not produce any rice buy it from the church, while people who hold cash even buy it from the market. This is an indicator of villagers' changing food habits linked to rice production and consumption.

After the construction of the church, the institutionalized organization of labour power based on age sets and the youth was developed increasing efficiency and produced much surplus. Although surplus production was difficult given the limited land resources and supply of flat lands for paddy expansion, it could only be increased by timing agricultural labour during the cropping season. One of my informants observed that in the olden days, people did not care much about their crops. Labourers were not sensitive to intensive crop management. Job's Tears and millet grew well in hardy conditions. Villagers were busy defending their settlements and were engaged in settling feuds. These days, paddy cultivation requires at least two to three organized weeding periods, just after transplantation, before a bumper harvest can be expected. Similarly, in the uplands the plantation of cash crops and horticulture requires much labour, time, dedication, care and an organized labour force that need to be disciplined, effective and timely.

The institutionalized production of labour also set the dilemma in prioritizing farming that was dependent on *jhum*. As more and more wet-paddy fields were opened, new exchange relations of labour distribution emerged. Labour was dearer in the paddy fields. As one farmer remarked during fieldwork, 'we cannot take labour to the paddy field without offering them meals that had good protein source. We have to feed them with meat along with rice'.[6] In the upland farms, this was not the case. Labourers never demanded meat; that was expensive and an additional burden to the household labour.

The timing of cropping has also shifted with the intensification of long-bean cultivation and paddy farming in the lowlands. A comparison with the past calendar shows that farmers have moved the maize sowing time to the month of March, and is harvested in the late July and early August, before the long beans can be transplanted on the soil. The early sowing of maize is guided by the importance of sowing long beans in a time when it has high commercial value (a single tin of *kholar* weighing 15 kilograms fetches Rs 450 when sold in the Tuensang market and Rs 350 in the village by 2009 rates). A few decades back, both long beans and wet-paddy rice were minor and luxury crops in the farmers' fields. Colonial records on the tours conducted by administrators also illustrate this fact. Today, the growing market demand of long beans and the ready cash it offers to cultivators has made it the most important commercial crop in the village.

The demand for cash is triggered by people's increasing household spending on commodities and services that were not central to Yimchunger life a couple of decades ago (tea, sugar, milk powder and children's tuition fees, school uniforms, stationery, transportation costs and other services, such as electricity). Rice, on the contrary, enjoys popularity because of the 'symbolic capital' attached to its production (see previous Chapters 4 and 5 for discussion). It is often acknowledged as the crop of the civilization. Both rice and long beans are successfully grown by following calendar schedules and accurate timing that have a very limited link with traditional time reckoning in Yimchunger society. They are primarily dependent on a calendric schedule procured from the Agricultural Department, which instructs farmers when to plant crops. Farmers depend on their tacit knowledge and years of experience experimenting with rice. Government subsidies for improving water supplies and providing new seeds have defined the proliferation of rice cultivation. Farmers experiment with government-supplied seeds for which they are dependent on the official state calendar and instructions in cultivation.

During my fieldwork, villagers complained that under various rural development schemes farmers received new crops and seeds and that they had no knowledge of how to cultivate them. With the proliferation of new seeds and saplings from the state, cultivators are becoming more accustomed to calendar time and manuals on cultivation. Here, the role of village elders as knowledge-keepers of seasons is fast eroding and is being taken up by the new generation of farmers, who visit the Agriculture Department staff and take instructions from them for cultivation. Not only government agencies but also private farms, NGOs and GONGOS (Government Operated NGOs) instruct new generation farmers on the benefits of experimenting with exotic crops. Farmers with whom I interacted in the village, mostly young and middle-aged, were enthusiastic to experiment with new crops as long as subsidies, seed supplies and cash were ensured.

During my stay in Leangkonger village, one of the *khel* elders, also its leader, had planted a bio-diesel crop in his farm. The *khel* elder had introduced the crop by forming a self-help group (SHG) with his *khel* households as members. The *khel* members were obliged by principles and mutual obligation to give their labour for the success of the group. The SHG has evolved as the new mantra of development. None of these activities would be possible without the successful timing of farm action in the field. Labour supply during cultivation has become central to crop production.

Since 2007, some farmers have started using hired bulls from neighbouring villages to cultivate their paddy fields. The use of the plough is further changing the time and labour required for growing wet rice. Community development and institutional state intervention to improve farming by introducing new crops is often blamed on the farmers for their lack of innovation to cope up with the demands of the market. The success of commercial farming depends on a close interplay of symbolic and social capital determined by time as perceived by individuals and as groups. In the study village, the transition from Job's Tears to rice farming has taken place, and one of the factors facilitating this change has been the organization of labour and farm relations by the church through its labour associations in the village.

Time and the Body

One of the most fundamental goals of Baptist evangelists who came to spread the message of the gospel in Yimchunger village was to improve the lives of the *jhum* cultivators through instilling the Baptist faith and moral standards that discredited a hedonistic lifestyle. The success of such moral standards was measured through abstinence from illicit copulation, the drinking of homemade beer and the renunciation of headhunting and inter-clan feuds. Headhunting died down with the extension of the post-independence administration and the growth of the missionary faith among the Yimchunger villagers. The missionaries became more tolerant, recognizing that their codes had little impact on sexual promiscuity and moral delinquency. More important was the broader structure of discipline, within which the goals of moral development could be achieved: it was not only leisure time that was to be regulated, but the minds and bodies of the people were to be regulated as well by creating disciplined bodies regulated by weekly routines of both work and leisure.

The missionaries valued promptness, cleanliness, hygiene, hourly routines and schedules not only as preconditions for wage-earning and employment, but (in the missionaries' minds) as ends in themselves – as correct ways of living. Protestant school teachers inculcated a pedagogical practice that cultivated in young children's minds – and bodies – timetables, school bells and examinations. Villagers who did not send their children to school were seen as backward. In the church youth gatherings and at Sunday service, educated youths would participate in 'exalting', uplifting sessions to inspire people with the goodness attached to getting educated and pursuing white-collar jobs and city life. In many inspiring guest speech sessions which I participated in, the village youths named their brave elders, who were now officers in the government and in public service. Through education, one gained higher status and prestige in the village as a patron – a knowledgeable soul who could understand the villagers' needs and aspirations. In the pre-Christian days, training had been through group participation and apprenticeship. In both government and religious schools, however, a multiple and progressive series of tasks was set, each followed by a test, and with students constantly supervised to keep them from being distracted from their exercises. Such discipline, according to Foucault (1977), opens up an analytical space that coerces not only bodies but also minds. In learning to obey the new routine of elementary schools, Yimchunger children also learnt to think about the church and officials in a new way.

Missionaries had instilled the ideal of improvement and new morals, in which working in mundane ways in the fields became difficult for the younger generation. They felt frustrated at their inability to maintain the standards they had been taught in schools. In recent years, there has been increased household spending on child education, sending them to missionary schools for quality education. This came out very clearly in my household survey.[7] A substantial portion of household income was now spent buying milk, sugar and tea leaves, as people substituted drinking tea for homemade beer, a beverage consistently opposed by the church. Pigs were no longer sacrificed at traditional feast times but to celebrate a child's admission into school and to cover their admission expense.

For the pre-Christian Yimchunger, time was valued and expressed in daily practice in terms of feasts and reciprocity (labour contribution to the village chief's family). One of the Yimchunger villagers who had travelled to my host's house narrated the value attached to the rearing of pigs and the sacrificing of *mithuns*. The older the *mithun* and the fatter the hog, the more it brought prestige to the feast organizer. Similarly, prestige was gained through 'feasts of merit', which involved enjoying rice beer and which lasted for many days during the harvest season festival. With the intrusion of the missionary faith, all such activities were tainted as 'economic waste'. But to the pre-Christians, it was wealth – a resource for symbolic capital. E. P. Thompson's famous work, *Time, Work-Discipline, and Industrial Capitalism* shows how the historical transition to industrial society in the West 'involved a restructuring of working habits accompanied by a new "time sense", which changed inward apprehensions as much as external activities' (1967: 57).

Thompson also describes how, in the new time sense, 'attention to time in labour depend[ed] in large degree upon the need for the synchronization of labour' (1967: 70). I have discussed above how a concern with the synchronization of agricultural activity was expressed by the *khiungpuh*, the highest ritual office among the Yimchungers, because of his control of time and kin labour. This new-style approach contrasts with the sense of time, as Hoskins has shown in the context of Kodi swiddeners in New Guinea, as not 'tied to the notion of value in a relative rigorous fashion; rather it was presented as characteristic of prestigious wealth objects and a measurable attribute of sacrificial animals' (1993: 219). The Ao and early Sema Baptist missionaries brought in new temporalities and meanings of time and discipline into Yimchunger life, based on a seven-day week with prescribed days of rest and worship, followed by work. The value of time, as connected with the value of life, in pre-Christian Yimchunger society contrasts strongly with Baptist time reckoning that, valuing their 'Protestant work ethic', organized time as toil and discipline for self-improvement.

Conclusion

The chapter has focused on the politics and the control of time in rural–agrarian Nagaland by the Baptist Church, as it shaped the political and social relations in a Naga village. By studying the dialectics of time and work imbibed through mission work, I have contended how Protestant work ethics permeated people's lives. In village studies in Northeast India, the church has been narrowly defined as a social institution of faith, as the regulator of people's moral behaviour and wellbeing. In this chapter, I have described the church organization of time as an object of power relations that controls social relations. With the diffusion of Christianity and the mission calendar in the late 1950s, the notion of time was re-imagined with the practice of observing Sunday as a rest day. The introduction of the mission calendar brought particular days of fasting, celebration (Father's Day and Mother's Day) and free man-days for the church that were distinct from the farmers' traditional ways of connecting with time. The church organization of agricultural time came through the use of a calendric schedule for cultivation. The ritual power of the chief clan was replaced by the church pastor and the new institution of missionary faith. In the next chapter, I discuss the NEPED project and how villagers rearticulated the meaning of *jhum* land development programme introduced by the project staff in the village.

7

Micro-Politics of Development Intervention: Village Patrons, Community Participation and the NEPED Project

In this chapter, I study the project text – programme plans, handbook and reports – produced by a development agency that has been in operation since 1995, the Nagaland Empowerment of People through Economic Development project (NEPED). A key feature of the approach followed by the project officials is people-centred participation in the project, drawing on the community's 'social capital' to act together as a community.

I examine the project outcome as a 'participant outsider': one who has had (limited) access to the project staff (planners and implementers) by drawing on beneficiaries' experiences through participant observation, a household survey and interviews, apprenticeship with a farmer to understand the art of cultivation and through spending considerable time in a post-project beneficiary village learning what people thought about the project and how they had participated in it. The project perspective is thus measured against the subjective experience of the beneficiaries (and non-beneficiaries) engaged in the development programme. The key focus here is the part of the NEPED project through which soft loans favourable to farmers were provided in a programme called the 'Revolving Fund' to improve farming in the village of Leangkonger located in Tuensang district of Nagaland. The chapter thus builds on the previous Chapters 5 and 6 where I discussed the questions of patronage and control over labour by village *khel* (ward-sector) leaders and the Baptist Church. I take a few loan beneficiaries as central figures to demonstrate how loans and relending operate within the emerging social structure of the village.

The chapter is divided into two parts. In the first part I discuss the NEPED project and in the second part, I engage with key project beneficiaries through listening to their life histories and how they received development grants and reshaped project goals in an attempt to give new meaning to the concepts

Image 7.1 Naga villagers as represented by the project staff.
Source: NEPED, 2007.

of the project's success or failure. This chapter contributes to the broader argument on how the agency of the actors who participate in swidden (*jhum*) improvement programmes should not be discounted when discussing development interventions. The villagers, who are represented in the project literature as 'objects of development' – as simple and poor beneficiaries – exercise their agency in ways that significantly shape project outcomes. In this discourse of farmers' development, the project villagers are represented in NEPED publications by emphasizing the outcome of implementation through sketches that depict poor marginal villagers as ignorant and aspiring for 'improvement' (Image 7.1). These sketches, which are powerful images of 'simple, ignorant villagers', as people in need of aid, admonishing their irrationality, but showing them responding to development, also influenced outcomes (Hobart, 1993). The pictorial description of beneficiaries as ignorant and simple displays a representational politics of backwardness and underdevelopment. It has clear parallels with the colonial ethnographic representation of Nagas through photographs as noble savages (Image 7.2).

Here, I also engage with the moral discourse of development that 'never fails,' as projects continue to build on their failures and successes through a discourse of 'improvement'. As David Mosse has argued, project success and

Image 7.2 Yimchunger Naga villagers drinking locally made wine. Photograph taken in Kutur village by anthropologist Christoph von Fürer-Haimendorf (1937). *Source*: SOAS Fürer-Haimendorf Digital Archives (JICS), PPMS19_6_NAGA_1070.

failure are more a practice of interpretation (2005: 18). I exemplify this case by looking at the NEPED intervention in my study village, Leangkonger, and thus see my analysis as adding to the arguments made by Mosse (2005) and also Leigh Pigg (2002) that community development projects have an inbuilt rationale of progress and moral improvement that is cast through the objectification of beneficiaries as 'primitive and backward' and as recipients of 'aid and subsidies' (see NEPED handbook 1999, 2007). In development projects that are technically driven by experts, elite 'indigenous knowledge' keepers act as brokers by influencing project outcomes that expose not only the ideologies of donor-driven development (top-down) approach but also make us think about the agency of individual beneficiaries who act as project recipients (Green 2002).

Following the case-study approach, I later show how donor evaluation practices like participatory rural appraisal (PRA)[1] mute emerging social inequalities and shape project practice. It is equally important to address the project's deployment of the 'community participation' rubric. I aim to problematize the ideal conception of participation and development through community collaboration by highlighting the different interest groups who benefit from these projects. 'The community' is, in reality comprised of social webs and networks of patronage and kin relations that shape project success for the beneficiaries but exclude others due to their inability to develop patronage with their fellow villagers. While such failure can be understood – as Appadurai has expressed in his study on slum dwellers – as the lack of these villagers' 'capacity to aspire' (Appadurai 2003), this lack is conditioned by the social circumstances and marginality produced not only by the local power relations that shape exclusion but also by the ideologies of governance that link communities with market and neo-liberal policies followed by national governments (Li 2007, 1996b). This, then, has implications for inter-household relations and social stratification among highland communities (Li 1996a).

The Project

The NEPED project began in 1995, as a *jhum* development and alternative livelihood regeneration programme based on community participation and farmers' local knowledge. It was supported by the Canadian International Development Research Centre (IDRC) and the India–Canada Environment Facility (ICEF). The Nagaland government was the partner–local body that oversaw the project by providing manpower and infrastructure to set up the office and by establishing the NEPED as a government-organized non-governmental organization (GONGO). There were other stakeholders in the project who, in different project phases, provided technical support and 'expert knowledge'.[2]

Leangkonger had already become a successful pre-project model village taken up for project implementation in Phases I and II (1995–2005). At the successful completion of those phases, the NEPED project staff prepared reports in which they highlighted: how participatory development at the project site was carried out; what tools were used and what practices were followed; what lessons were learned; and what outcomes resulted (NEPED 1999, 2007). The project's ultimate goal was to bring economic and social development through the diversification of subsistence farming. It was assumed that the introduction of cash crops, soft loans and the formation of self-help groups would lead to settled cultivation and empowerment and replace the practice of 'slash and burn' with sustainable agriculture.

The NEPED project had two phases. The first lasted for six years, from 1995 to 2001. During this period, the project officials claimed the planting of six million trees in *jhum* fields covering 854 villages (NEPED 1999). The thrust of the project shifted to provide better livelihood through economic empowerment programmes like micro-credit and the formation of self-help groups during the second phase. The idea of micro-credit was experimented with in the form of a revolving fund (loans made available to farmers) that was introduced to wean people away from a 'subsidy culture' and to introduce agricultural activities such as the cultivation of cash crops (like ginger, cardamom, passion fruit, chillies and other horticultural crops), under shade trees and as creepers (NEPED 1999: 99–123). The fund was rotated as a loan with a soft rate of interest and was managed by village council and development board members.[3] With every new phase, the project came up with new ideas and agendas for improvement as the funding priorities and project planning shifted to create new opportunities for sustainable development.

The NEPED project was one among many such programmes designed in the past that were constituted with an elaborate administrative structure. The programme was implemented through a decentralized administrative setup with support and cooperation from the village council, the development boards who worked in liaison with Project Operation Unit (POU) members, team leaders and village-level facilitators of community conservation (FCC). The latter were NEPED staff selected from the beneficiary villages. The project heightened people's expectations of more loans and cash as disposable income, which the beneficiaries rationally used to satisfy pressing family needs rather than farm improvement. It was only in a few plots that farmers experimented with the introduced crops. The crops were successfully grown by only a few farmers, who received constant support from the agency. The NEPED team built market links with town merchants who had enough food crops and farming plots to satisfy their family needs. In the fields of other farmers, experimentation with new plantation crops was associated with crop failure. Many used NEPED's Revolving Fund (soft-loan scheme) for nonfarm activity and expected to recover this loan at the end of the five-year period through the interest they would earn by reloaning these funds to fellow villagers and traders who could not become beneficiaries as they lacked the goodwill with the village elites. The consequences of these initiatives will be discussed later in the chapter.

The NEPED Model Village

During my first visit to Leangkonger in 2009, I was not aware that it was a model village presented by the POU team as a success story in the transition from *jhum* farming to more sustainable relay cropping – a form of farming in which leguminous crops such as *kholar*, French, or long beans, are grown in combination with soil-exhaustive crops like corn so that the soil does not lose its fertility. (See Image 7.3 on the Leangkongers' success story with *kholar*.)

As a result of the POU report (see Image 7.3), the village became a recipient of a major development grant during the second phase of the NEPED project (2001–6). The village was already a success long before any development

Transition Story: Leangkunger's Local 'French' Connection

Leangkunger Village in Tuensang District is located in a remote corner of Nagaland bordering Myanmar. Although the farmers were traditional shifting cultivators growing maize and paddy, their primary crop has now become the French beans (locally known as *kholar).* The beans were generally grown for home consumption but it spread through occasional visitors or relatives to other areas. The increasing demand from other areas for *kholar* made the farmers realise that their humble backyard French bean variety was a hot marketable commodity. They increased their crop area under this variety.

The demand for the beans increased the interest in *kholar* and soy bean cultivation in Tuensang District thus leading to diversification of crops and the gradual transition from subsistence to cash crop farming. Encouraged by their success with *kholar,* farmers have also taken up maize and local potato cultivation in large areas.

While the initial push for transition from subsistence to market-based farming was because of demand for local crop varieties, the rapid transition and shift towards cash cropping came as a result of project interventions (revolving funds, marketing know-how and raised awareness about marketing). The Government of Nagaland, after seeing the success of this local French Beans (*kholar*), declared the entire Tuensang District as a "*kholar*" district.

Image 7.3 NEPED handbook showing Leangkonger success story.
Source: NEPED, 2007.

intervention had taken place. However, during the third phase (2008–12), funding was withdrawn as project officials on the ground became aware that their loan grants had not made a significant impact and failed to bring about the expected results from the relending of the revolving fund among beneficiaries. Villagers reported that the POU members who monitored the programme were not cooperating and unwilling to visit the village at the end of the project's phase two. The POU, on the other hand, realized that the farmers had failed to achieve the desired goals and were not repaying the loans; nor in many cases had they succeeded in growing cardamom (introduced as a cash crop) for the market, thus failing the project mandate.

The introduced incentive to grow ginger as a cash crop also failed, while cardamom was successfully grown by a few farmers who were nonetheless finding difficulties in protecting their crop from the vagaries of weather and fluctuating market price. The 2009 dry season almost wiped out the remaining cardamom plants. When interviewed, farmers reported that they were 'lazy' to grow cardamom, as they had more important food crops to sow and harvest during the crop season. The incentive from the market was low, due to the weak link established between the farm and the place of sale. Only farmers who regularly visited the towns and had built good networks with local merchants sold a few bags of cardamom annually at very low prices, as they had very little opportunity to bargain. Another aspect of the narrow incentive to grow cardamom that farmers cited was the revolving fund money, intended to purchase seeds, which they received as part of the loan package. The uncertainties of cardamom crop production had made many beneficiary families default on their loans, including my interlocutors and village headman. The project officials branded the villagers 'lazy and corrupt', as was revealed by farmers, saying this was the reason for the demise of the revolving fund and of the cardamom plantations. This was how the NEPED project was phased out in the village.

The picture as presented by recipients and donors is complicated. During the fieldwork, I was able to discover why cardamom plantations failed when the farmers had been successfully growing long beans for many years. Indeed, it was based on this success that the village had been selected for cardamom and ginger plantation. Twenty families were identified as beneficiaries in the cardamom plantation programme during (2003–7). However, only a few had been able to continue the project on their individual farms by the end of the project phase. The village agrarian economy was dominated by relay cropping of maize and long beans, while there was no trace of the newly introduced crops except in a few plots where farmers struggled to grow and maintain their cardamom plants. The failure, later discovered, was due to planners' 'expert' understanding that had discounted farmers' needs and local networks of demand and supply. Farmers cared least for the new crop, cardamom, as they found it hard to grow, nurture and care for once there was no incentive in the form of loans. They had no previous experience of cultivating and selling cardamom. The vagaries of weather also contributed, turning initial experiments into failure in many plots as farmers complained of a high sunshine factor, implying there was no shade to protect the young cardamom plants. The NEPED soft-loan scheme that placed its faith in the revolving fund automatically suffered a setback. The POU members who were excited by the villager's innovative crop rotation with long beans as a cash crop soon realized that their recipients of aid had mismanaged the loans by not using them efficiently. This led to the retrenchment of support and a subsequent loss of interest on the part of the farmers, who were struggling to increase

Image 7.4 Maize and long beans relay cropping in Leangkonger.
Source: Author's photograph, 2009.

the production of the long beans in their *jhum* plots to maximize their cash income.

Leangkonger is located at an elevation of 1300 m above mean sea level and was already popular in the district as one of the best long bean producers among the Yimchunger villages (see Image 7.4). Long-bean production has revolutionized farming and, in recent years, has become one of the staple food crops of the village, replacing Job's Tears (buckwheat) and millet from their diet. The intensification of long-bean cultivation and relay farming of maize has established a healthy *jhum* cycle in several farms where farmers can now work a single plot of land for several years without shifting their fields. Fields close to the village settlements were already becoming permanent sites of cultivation with multiple agro-horticultural plantations and belonged to powerful households in the villages who were the beneficiaries of multiple rural agrarian diversification programmes.

In Leangkonger, the rural *kulaks* are the village headmen and political party intermediaries who claim to be the *khel* heads. The council registered 198 households for Mahatma Gandhi National Rural Employment Guarantee Scheme (MGNREGA) man days – while, according to the electoral roll, there were only 159 households in the village. In my study, I recorded only 140 households when I carried out my baseline survey, of which 106 households participated in my interviews.[4] The variation in household numbers is based on the manipulation of figures while reporting to state officials on the family size of households. I concurred from my fieldwork that the village was divided for the MGNREGA man days programme into nine units (wage groups), each of which had a *khel* head and a *khel* secretary, who were responsible for overseeing and ensuring the participation of their *khel* households.

The Project Operation in the Village: Patronage and Privileges

The Project Operation Unit staff who initiated the programme in Leangkonger was related to the village, as he had married one of the women from the *jangrü* clan. His father-in-law, Antham,[5] was the link in the village and responsible for the implementation of the project. Antham, as I discovered while doing fieldwork, had a vibrant history: in the 1960s, he had lived in a neighbouring village, served as village pastor, and later joined the Naga National Movement for a few years and went underground. Dissatisfied with the struggle for independence, he returned to his village. Then, after becoming a government primary school teacher, he was posted to Leangkanger. After a few years of service, he took voluntary retirement and became a member of the village citizen group. In the past decade, he had been involved with the rural development department and currently acts as the manager in charge of the distribution of corrugated galvanized iron (CGI) sheets in the village under the Indira Awaas Yojana (IAY) government housing scheme. His children (nine daughters and three sons) were all grown up and had mostly migrated to the cities and towns. One of his daughters was married to a subdivisional officer (SDO) in Kohima and this son-in-law was the person in charge of implementing the project in Leangkonger. The officer visited his father-in-law's home annually and was himself a Yimchunger Naga and knew the village well.

The NEPED project planning unit entrusted Antham with the responsibility to encourage cardamom plantation and identify beneficiaries. In his interviews he proudly claimed that the cardamom project was initiated by him in the village. The loan offered to each project beneficiary family was equivalent to £250. The council, along with the development board members, selected 20 beneficiaries. Among them were three *gaon burhas*, Antham, the former village pastor, and the village guard commander, and they were among those who were also beneficiaries of the project. Before the seeds were dispersed, the villagers were given training in Zunheboto district on how to plant and grow cardamom. The subsidies of £250 were given to buy seeds, manure and farm implements. Antham's farm became the test plot and focal point for dissemination.

My interviews with the beneficiaries revealed that the crops had failed in their individual plots. The plants survived in only two plots, Antham's and in a *gaon burha's* plot. In 2009, the lack of rain meant a poor cardamom crop. Equally, farmers like Antham could not sell the cardamom even in Tuensang market, which was some 65 kilometres from the village. The limited production also meant that it was not profitable to sell the produce in such distant markets. The NEPED POU staff who oversaw the project complained that the farmers were not as enthusiastic in the cultivation of cardamom as they were in the cultivation of long beans. They further clarified in my interviews that the villagers' lack of attention and care for the crops had potentially inhibited the project. These remarks were not different from what NEPED bosses would tell me in their headquarters in Kohima, the state capital. For every success story, the NEPED officials took full credit for their intervention while failures were blamed on farmer's laziness, corruption and disinterest towards improvement.

The POU members in their interviews did not reveal who the beneficiaries of the cardamom project were and how they were selected. The micro-politics of selection is central to understanding project failure. For any mode of farming to succeed, labour contribution is required. Interviews with the beneficiaries revealed that their interest in the plantation was due to the (diverse) circumstances and conditions in which they were living and not necessarily for farm improvement and diversification of livelihood. For the former village pastor, gaining access to a NEPED loan was a means to redeem his debt and to procure food for his family. For the village guard commander, these were added funds to invest in new property in the village, while for others, the money was important to meet their immediate household needs and for investment in tangible assets, like buying agricultural plots, houses and hunting guns. Those who had genuinely invested in the cultivation of cardamom could not maintain their fields, as there was no immediate return in the first year and it was thus too costly to devote the labour that could be directed to staple food crop production. For the *gaon burha* and Antham, their role as brokers had facilitated the patronage they had built with clan and lineage members, who, in return for becoming beneficiaries were obliged to offer free labour to them. Their social position in the village as brokers in development projects had many supporters who were ready to lend help in return for subsidies and financial benefits. Brokerage entailed both the material and symbolic benefits they enjoyed with villagers as experienced in other parts of South Asia (see, e.g., Mosse and Lewis 2005).[6] By the end of the project's second phase in 2006, the cardamom plantations in the village had been reduced to a few plots of land.

The beneficiaries also complained in their interviews that the POU staff were not very happy with the villagers at the end of the project's second phase due to the project's poor performance. The POU restricted their visits to the village after 2007, when the final review of the second phase was completed. The last time they visited the village was in 2009, when I was in the field. This time the POU staff came with a grand new idea of conserving biodiversity through spreading awareness of endangered species among villagers and making provisions for their conservation.

The use of the revolving fund provides important insights into how funds are managed by village beneficiaries who are also responsible for the implementation of the programme along with NEPED staff and their village coordinators.

The Predicament of Project Implementation and Its Outcome

The household survey I conducted was completed among 106 families in Leangkonger village. Based on the survey, I marked the household in each *khel*. Although the number of total households in the village is a little higher, I could not cover all of them as, in some families, people were living in the nearby town and seldom visited the village. According to the sample survey, there were 140 households in the village, divided into nine *khels*. Leangkonger had five clans, among whom the *jangrü* clan had the largest number of households. The clan division was an important aspect of identity formation and group affiliation. The numerically bigger *jangrü* clan had four divisions within them and seven *gaon burhas* who were members of the village council. Similarly, the *jangrü* clan members had within them political party workers, government servants who acted as go-betweens in development projects. During elections the *jangrü* clan played a decisive role as they were numerically the largest. Anyone who could garner the *jangrü* vote would win the election. The village had two grocery stores and several small *paan* (betel leaf) shops, where commodities like cigarettes and tea, biscuits and milk powder were sold. The village also had two Nepali heads of households who are now considered to be Yimchunger after marrying village women.

There were many brokers in the village who were government employees or political party actors. They often contested the traditional authority of the village headman and influenced their actions, as they knew the rules of governance and public policy on rural development, so they had established patronage over the village. They acted as influential people and held responsible posts such as council chairman. However, these posts were highly contested and abused by influential men in power. During fieldwork, the post was held by the second *gaun burha*, as the person who formerly held the post was charged with corruption. The NEPED project funds came to the village council through the SDO father-in-law who was also the village facilitator for the CGI sheet scheme under the Indira Awaas Yojana. He informed me in confidence that the villagers are ignorant, illiterate and poor and can be easily manipulated. This disadvantaged them greatly compared to village elites and government servants, who knew the bureaucracies and could easily benefit from government programmes as beneficiaries.

According to my baseline household survey, 28 households were beneficiaries in the NEPED project during the various project phases (I and II). All these villagers got revolving fund micro-credit loans of between Rs 5,000 and Rs 20,000, to be repaid at the end of the term with an annual interest rate of 5 per cent. The church pastor was one of the beneficiaries. He received Rs 100,000 – 10 per cent of the total grant. The villagers undertook the revolving fund scheme in an innovative way. The pastor reloaned the money to town businessmen and some Muslim shopkeepers in the village at a 10 per cent compound rate of interest per month (120 per cent annually). It proved to be a failure, as beneficiaries who were reloaned monies could not repay such high interest. The pastor, in his interviews, reflected that he was not interested in reloaning the money among villagers as they did not have surplus cash income at hand and would soon default. Many of his borrowers (mostly migrant grocery store owners and petty businessman) ran away with the loan money and never returned to the village. At the end of the project, the pastor was unable to repay even the principal amount. However, his social position as a religious head meant that other beneficiaries did not dare question his integrity. His social capital and support within the village council as one who would rescue villagers in need further established his social position and prestige.

I now present some household case studies to show how the fund was utilized by the NEPED project beneficiaries as a measure to materialize well-being and social protection through micro-finance. It must also be highlighted that 48 households, nearly 50 per cent of the village, did not receive any funds from any of the government programmes running during my fieldwork, and hence were excluded from loans and subsidies.[7] Public service employees were denied loans on the grounds that they already had a fixed source of income and also were eligible for loans from their service department. However, there were exceptions. The person through whom the NEPED cardamom plantation subsidy was received in the village was a government employee, but because he facilitated the whole process (as a go-between), he became a beneficiary in the cardamom plantation project.

Household No. 5 in the upper *khel* was selected as one of the beneficiaries in 2000.[8] He received Rs 10,000 as loan money and used it for his immediate family to build a house. He could only repay the loan with interest in 2008. The NEPED project ended within five years, but he, like many beneficiaries in the programme, was unable to repay the loan on time. The revolving fund, based on interest generated out of the initial allotment fund, never materialized in practice. In order to get these loans, people had to show collateral in land and household assets. Despite failing to repay the loan money on time, the households were lucky not to have faced confiscation of their property — they were in good terms with the council members. For many farmers, the loan money was used as a resource to build new assets, buy land and send their children to better schools.

The overall mandate of the project was therefore reworked by the villagers through their individual and pressing financial needs. This revealed how project policy and practice are influenced by recipients and how unhelpful it often can be to divide the two actors in aid-driven community development. As Rossi (2006) emphasized, we need to think less in terms of interfaces between separate worldviews and more in terms of positioned strategies and perspectives while thinking in terms of donors and recipients in development programmes. Some of the household heads (Household No. 13, for example) were not in a position to receive NEPED loans as they had limited collateral and very little social capital to convince the council that they could repay the loan amount. Farmers like my host, Nikon,[9] did not bother to associate with the programme as it was managed by another influential villager who had all the resources to mobilize labour and NEPED benefits to his own farm. Instead, Nikon nominated his brother, the village commander and his sister's husband, an ex village pastor, as beneficiaries in the NEPED programme. With this, he established his social position as a benefactor and strengthened kinship ties through his blood relations.

For the cardamom plantation project, the village council formed a group of 20 farmers to be beneficiaries under the NEPED scheme. Thirteen farmers received a sum of Rs 20,000 while the remaining received Rs 10,000 each. For two years, from 2006 onward, Household No. 17 sold cardamom and made a profit. In 2009, due to bad weather, all the cardamom bushes were destroyed by strong sunshine and there was no prospect of growing them anymore, even for the most prosperous farmers who had maintained the plots with the NEPED loan and farm labour offered by the beneficiaries. The cardamom that was produced was sold to a local merchant from Kerala who bought them at Rs 55 per kilogram. Household No. 17 was able to sustain the cardamom plantation, as his farm was selected as a test plot. Farmers who joined the project were obliged to work on his farm during the initial period in order to receive seedlings from the test plot. They devoted a portion of their weekly work towards the maintenance of cardamom. Once the bushes were ready for transplantation in the field, the loan beneficiaries cooperated with the household in his farm. The obligation was ultimately only one-way, as the beneficiaries could only get the saplings when they participated in Antham's test plot. Once the saplings were transplanted by individual participants into their own farms, they failed to take care of them as they provided only limited marketing opportunities. Their own farm work for food crop production was more important than these new plantations, which required enormous attention and often conflicted with their own crop cycle and the labour supply in the village. With the subsidies, they were encouraged to maintain the plantations. Once the NEPED fund was distributed, and there were no avenues for fresh subsidies, farmers stopped bothering with their cardamom plantations. The expected produce also fell and few individual households took the initiative to sell their produce in the market. These outcomes disappointed the POU staff, who claimed the villagers were 'lazy' and had misappropriated the funds. For the next phase, the NEPED project staff did not shortlist Leangkonger as a project village, even though before the start of the programme, the village had been presented as a success story given its transition from *jhum*-based millet and Job's Tears cultivation to long-bean cash cropping.

Similarly, rural elites have access to the towns and they constantly bring new ideas, luring villagers into new schemes and programmes for agricultural development. For Household No. 21, a NEPED loan was out of the question as they were too old to work hard or invest in agriculture that would generate the repayment of a loan. As one farmer observed, only households that had established trust with the village council members and those who were engaged in village politics had the credentials to get a NEPED loan. The loan distribution was also linked to the personal affiliation of village members and their status within the church. Deacons who were very close to the church, village council and development board members and political intermediaries, the most affluent and influential men in the village, managed to obtain these loans and become registered as beneficiaries.

Household No. 40 had enough resources for subsistence. They received a loan amount of Rs 10,000 and used the money for paying the tuition fees of their child who was studying in a Catholic school in a nearby town that was considered to be better than the Baptist school. They valued spending the loan money to provide better education to their children. At the end of the loan period in 2007, they had to sacrifice domesticated animals to pay off the loan amount, which they felt was hard but inevitable to save their agricultural plot from being confiscated by the council.

From the household survey as described above, it can be inferred that individual households strategized their livelihood priorities with respect to the NEPED revolving fund scheme that was primarily advanced to improve livelihoods through cash-cropping activities. While a few farmers utilized the loan for agricultural purposes, not all were successful in cultivating cardamom and ginger. For other farmers, building houses, paying for their children's education and investing in more tangible assets were far more important than investing in farming. Other loan recipients used the money to develop innovative strategies of patronage with fellow villagers: redeploying the rationality of credit (loans), they became village moneylenders. Though unsuccessful in recovering the loan money, they clearly showed how cash can be productively reinvested to earn a speedy profit. The net result was that investment in cash-crop cultivation became the farmers' last choice. Only beneficiaries who owned multiple agricultural plots and were well connected to the market invested the revolving loan amount in cardamom plantation. The uncertainty of the monsoon rains, harsh weather conditions and the low incentive from the market meant that the initial interest waned. By the end of 2007, POU members from the NEPED department ceased making regular visits to the village. The project outcomes were far below their expectations. Many of the loan beneficiaries became loan defaulters as they did not invest the money in productive assets that could generate income. The NEPED officials had nothing left to promote in the village. During that period, the NEPED Phase II had also come to an end and the project staff were gearing up for the third phase, which aimed at a different set of empowerment and livelihood options centred around the conservation and maintenance of biodiversity.

The important questions that arise out of these case studies reveal how the idealized notions of 'community' and 'social capital' is mediated by identities, interests and factions within a project site. It is also important to recognize how the community actors, beneficiaries and project recipients build their relationship and participate, often rationalizing decisions based on their family needs rather than following the broad policy goals of donors or project planners. While project discourse may describe them as primitive, lazy and timeless rural folk, the project beneficiary community presents a more complex history and contains power hierarchies that are far more intricate than the simplistic fixed images of idealized tribes and homogeneous non-stratified villagers. The idealized utopia has rationalized the principles of community participation and commoditization programmes in Nagaland.

This draws attention to what Li (1996b: 502) has highlighted in her study on Indonesian hill peoples: that struggles over resources are also 'struggles over meaning' that shape project outcomes and help us understand 'agency' and strategies of grassroots actors in development. Programme interventions based on 'rapid appraisal', as were followed in the NEPED project evaluation repeatedly at the end of each project phase, produced lists of typical tasks, assets, outcomes, learning and needs, but missed the subtleties of the relationships and processes that define projects. They failed to document the strategies adopted by the individual households that determined the project outcomes. While farmers enthusiastically grew long beans, this did not automatically translate into willingness to adopt cardamom or ginger as a cash crop to improve household income. The physical isolation, distance and remoteness also made Leangkonger challenging for agribusiness. Nonetheless, the failure to grow cardamom had less to do with the sophistication of commercial farming and more to do with the project planner's failure to understand the communities' interests, motivation and strategies of action. These contingencies add to the picture of the 'high managerialism' of project practice that simplifies communities and their 'capital' (Van Ufford et al. 2006).

The fact that some people's actions constitute other peoples' constraints means that one cannot study individual behaviour in isolation. Or, in the language of economic analysis, one cannot ignore externalities. People's behaviours depend not only on what others are doing but also on the form and quality of the social relationships among them (Berry 1980). Farmers use agricultural loans for ceremonial expenditure in part because such expenditures serve to affirm or enhance their commitment to principles of interaction – among kinsmen, neighbours (Parker 1972). These principles, in turn, shape farmers' access to productive resources. Similarly, I have shown in my analysis how farmers reloaned the soft-loan funds to kinsmen who could not become recipients but were in desperate need of cash. Thus, denial and access to resources were shaped by the unequal power and patronage enjoyed by individuals within the community, which also shaped project outcomes.

The other point that demands attention here is the role of the brokers and intermediaries who became involved in implementing the project. These brokers worked both within the development institutions and the project village. The intermediaries played a very important role in producing a coherent development idea. The development programme staff, as seen in my case study, intervened in the field as brokers attached to their beneficiaries and the project community. Similarly, village brokers who took part as mediators express a unique relationship. As I have shown, the NEPED project came to the village with the help of the Block Development Officer who acted as the POU member and was the son-in-law of an influential villager. His relationship to the village was more than that of a project professional. Similarly, other project officials who worked for the agency were from the same community and often citizens of the village. Project staff members thus had a multifaceted relationship with the beneficiaries. They also shared unique relations of power and patronage that made reciprocation obligatory. In Leangkonger the end of the second phase also marked the termination of the Block Development Officer as a POU member, and he was recalled from deputation duties by the NEPED team. With a new POU member who came from a different district, villagers found cooperation less obligatory. Thus, development brokerage plays a critical role in shaping interpersonal relationships between villagers and project professionals who were also presenting communities as 'backward' in their day-to-day talk.

Every farmer who could secure a NEPED grant had his own life history of utilizing the revolving fund during NEPED phase II. Only farmers who had networked with the village institution – the church – and were members of the village council managed to become beneficiaries. However, the selection and distribution of grants were influenced by the decision of the village headmen and council that made these decisions based on which people they believed could eventually pay back the money. The council members informed during my interview that large grants could only be given to farmers who could provide collateral and whom the council felt would be in a position to return the money. The risk-taking farmers in the village were council and development board members who had both access to government programmes and land to show collateral for grants. This was the case even despite the fact that the NEPED action plan was based on participatory development. The bulk of the revolving fund entered non-productive services and ended up being seen as additional cash in the hands of villagers who were wealthy enough to manage their farms.

The distribution of the funds clearly showed that people who were close to the church, who owned multiple plots of land and who shared convivial relationships with the village headmen and council members received the loans. The outcome mappers 'project reviewers' hired as consultants were not in a position to take these factors into account when evaluating the project. The main measure used by the NEPED officials was the number of beneficiaries and not how the beneficiaries were selected or how the funds got allocated. They were engaging in narrating project success by describing the success story of relay cropping in Leangkonger as a positive precondition for development intervention. The NEPED revolving fund thus created new avenues of resource appropriation.

Project Outcome and Village Realities

Marley D. Faminow et al. (2001), who served as the director of the NEPED project from 1998 to 2000, applauded the project success in a paper he published by statistically proving that the number of plantations carried out in Nagaland initially in test plots were later scaled up. This was presented as a measure of the success of the first project phase (1995–2001). The results were, as the authors put it, 'extrapolated' from project villages and they presented a sanguine picture of the project success (Faminow 2001: 219). My fieldwork interviews with project beneficiaries in Leangkonger revealed that, in most cases, village elites described existing forest areas as newly planted forest when the project officials came to notify the test plots. Some farmers who had planted teak and gamari (Gmelina Arborea) a decade earlier under a different government programme also presented these plots as part of the NEPED success story of replication and intensification. Similarly, as the project report authors themselves recognize, farmers ended up planting and adopting commercially profitable trees as opposed to indigenous ones, which they put to everyday use as firewood and for building houses. Their choices were also influenced by the project emphasis on agroforestry and the planting of commercial crops and the likely economic benefits associated with timber tree plantations. On the downside, such rampant forestation has also made planted forestry vulnerable to forest fires, and many planters in Leangkonger reported that their plantations were destroyed by *jhum* fire.

Jhum fire is a common event in shifting agricultural plots during the dry season. However, in areas where hardy varieties resistant to fire exist, the forest fires cause less damage to tall trees and young stumps. In the case of teak, gamari and other timber trees, the risk of fire is enormously increased. These risks had been earlier assed by the Imperial Forestry Department reports. For example, writing on the need for terracing and pollarding in the shifting cultivation upland areas of Assam, the forest officer writes in the Agricultural Bulletin, published in 1938:

> The question of afforesting the *jhummed* areas by quick growing trees is a difficult one, because although there is a number of quick growing species, nearly all of them are more or less fire-tinder. This naturally limits the selection. If the danger from fire could be eliminated the number of varieties suggested below could be considerably increased. (Agricultural Bulletin 1938, p. 6)

These dilemmas had confronted colonial policymakers and also persisted in the NEPED project. When the project called for indigenous ecological knowledge of farmers and grassroots intervention, it was primarily focused on commercializing agriculture for the farmers and linking it to the market. The emphasis on timber tree plantation in the first phase of the NEPED project was influenced both by the promotion of non-indigenous varieties and the project's emphasis on timber harvesting in the long run as a profitable income-generation activity with *jhum* crops. Farmers found it illogical to systematically plant trees that already existed naturally in the village forest and were of less commercial value. Similar was their attitude to the *Nepalese Alder* that already grew in the wild. Farmers were aware of its quality as an anti-erosion, soil-protective, soil-regenerative tree. But they were reluctant to plant it as it required careful pollarding every *jhum* season and necessitated a long gestation period before any crop could be grown. Habitually *Nepalese Alders* created shade for the villager's primary food crop and acted as a hindrance to competing crops. Thus, alternative timber tree plantations that promised to generate cash in a few years attracted the beneficiaries' attention.

Much like the technical failure of the project to reconcile with timber plantation, the project officials failed to understand the social networks through which people established their patronage in the village. Individuals who were in power in the village council and worked as political intermediaries in the village networked the 'social capital' of the farmers through brokerage. When the project money was floated in the village, kinship and patronage relations became important in the identification of beneficiaries. The funds were initially distributed among the close-knit council members, while the leftover amounts was shared among members who had close ties with the council and village headmen. There was even a wide disparity in the *khel*-wise distribution of money. The *jangrü* clan members in the middle and lower *khels* who acted as political elites were the major project beneficiaries and, in effect, represented the 'beneficiary community'. The rest of the village could only get access to the NEPED revolving fund at a higher rate of interest through reloaning. The equitable distribution of project benefits and the identification of the poorest of the poor only remained on paper. The landless farmers lost out in the competition to secure loans, as they neither had the collateral nor the good will of the council. Most of the landless farmers complained in their interviews that the state benefit subsidies, loans and development schemes are first shared by the village elites who represent the village council. The patron–client relations are based on reward sharing between village elites and their political patrons.

In the NEPED project, the project officials' POU team used the pre-project success of farmers in the cultivation of maize and long-bean–based relay cropping as a successful example of agrarian change and improvement over the traditional swidden system that was widely understood as damaging to the environment. The village was presented as a case study of success in the programme literature for major intervention through development grants. As the villagers we interviewed noted, participatory rural appraisal techniques and village resource mapping were carried out to identify the nature of intervention. The project was also conceived to be carried out in a decentralized fashion, by closely working with village institutions and empowering them to distribute grants and select beneficiaries. The project outcome has been dismal. The formal equality mandated by the project does not work in the face of people's everyday knowledge that decisions are usually made by a handful of village members who have cultivated links with government and development professionals (see also Baviskar 2004). The Yimchunger Naga lineage-based social systems are networked with clan politics that determine people's choices and social behaviour. The clan heads and village elites who have control over their clan members manipulate and determine the success of development programmes. This is because clan leaders control labour relations that bring agricultural success and contributions to the village church. The patronage politics is oriented towards maintaining such relations through the identification of beneficiaries.

The project money was appropriated by people in power, and they manipulated and appropriated the project goal of the recirculation of cash through private lending at a higher rate of interest to villagers who were denied access to the project funds. The client list control over resources at the village level presents an important challenge to the desired goals of participation. Social capital is built on the networks of *khel* ringleaders who manipulate local patronage politics and determine the authority structure in Yimchunger villages. Collectively, they work with government staff and project officials to produce illusory success stories by emphasizing the circulation of grants without really making known how much of the population benefited by the programme.

For the NEPED team members and their funding bodies, project coherence as 'sustainable, livelihood redevelopment and capacity building' through participation and farmers' 'social capital' was a coherent idea that was to be maintained for future funding. The village beneficiaries tailored their needs and aspirations to receive opportunities for subsidies, loans and grants to build upon their existing personal wealth and patronage. The object of development was thus shaped both ways by the ideological monologue of the funders and the individual subjectivities of beneficiaries, who presented their 'will to improve' and enrolling themselves as beneficiaries. The various disjunctures that emerged in project policy and practice were concealed, subsumed and overlooked. The end of the first phase (1994–99) of the NEPED project was less satisfactory than the evaluators claimed. However, a powerful lobby by A. M. Gokhlah (senior IAS officer) in the government helped the project get the extended funding for the second consecutive term from IDRC–ICEF, that rarely supported projects for a second term in other project sites in South Asia. The continued support in the second phase with a revamped empowerment and commoditization objective became the touchstone of project success.

At the end of the second phase, much of the revolving fund had been dissipated within the community, defeating the very purpose for which it was created, thus ensuring the failure of the soft-loan scheme within the beneficiary community. The NEPED project staff had found biodiversity, indigenous knowledge on medicinal plants as a new object for project intervention. Parallel to this, a mini-hydel project based on local renewable energy came up as an additional input for a sustainable livelihood strategy and community participation.[10] The Ministry of Rural Development replaced IDRC–ICEF funding to rescue the NEPED programme in the third phase. Project staff complained that they had to follow the ministry's directive of one-third resources allocated to afforestation, which the project planners felt was unrealistic and a hard stipulation to implement given villagers' attachment to *jhum*. During the last days of my fieldwork in 2009, the NEPED officials visited the study area flying in on a helicopter with the German Development Bank staff. After further inquiry, I realized that the NEPED project had now entered a new era of funding that prioritized increasing the area under forest cover through the documentation of biodiversity and local knowledge about medicinal plants that would help conservation. This time around, the Asian Development Bank and the German Agency for Technical Development had endorsed the project with financial support that would help NEPED staff work on new areas of community development through people's participation. Thus, in the project planners' discourse, the project never failed, but in each project cycle, its failures and gains were shifted by new priorities and objectives that were set out by the funders.

8

Conclusion

I began this book by narrating the binary concept of how people narrate *jhum* as a non-sustainable wasteful practice, and at the same time see it as central to their culture. I return to this proposal in the conclusion by drawing on the histories of received wisdom and local innovations that have brought changes in swiddeners' lives. I draw upon the principal arguments developed in the previous chapters, organizing this summary around themes: state–Naga relations, episodes of violence, *jhum* conservation, state science and patronage, state and missionary mediation and NEPED intervention that have shaped the development discourse surrounding *jhum* settlement in the Naga Hills and among the Yimchungers in particular.

I mentioned in Chapter 2 that the monograph was originally designed to work on a *jhum* development programme, the NEPED project, its policy and practice of *jhumland* development in Nagaland. I was made to believe reading the NEPED project literature that their programme was a unique intervention, where in the past state intervention aimed prima facie at eradicating the *jhum*. The NEPED project claimed to have reversed this approach through valuing indigenous knowledge of *jhummias* and promoting participatory planning and development by putting local farmers at the centre of *jhumland* improvement. However, my relationship with the project officials came to a crescendo only after a few months of fieldwork, and this resulted in complete inaccessibility to the inner world of project staff and consultants. I was put off this aspect of the study for good, and this turned out to be a new challenge and at the same time an opportunity to look at the wider timeline and social field of development in contemporary Nagaland by tracing development back to colonial times and unpacking Naga–state relations.

I based my fieldwork in a Yimchunger village after a long struggle but, quite unconventional to anthropological tradition, I moved back and forth between the village, the town and local archives, making sense of the longitudinal history of colonial contact and post-independence intervention in the Naga Hills. I thus referred to the rich archival documentation that became the source of my understanding of colonial state control alongside the numerous monographs that were written in the 'Notes and Queries' style of ethnography. Gleaning colonial archives helped me to avoid writing another synchronic 'history-free' detailed account of Naga material culture, social organization, folk stories, kinship relations and agriculture.

I also explored a wealth of information in the visual tradition that opened up a new genre of looking at the power exercised by the colonial administration through visual tropes of the Nagas as 'primitives' and 'noble savages'. I found resonance and a continuation of this visual tradition in government literature, missionary magazines, souvenirs and, more recently, in a variety of visual ethnographic research carried out in the Naga Hills by scholars like Ganguli (1984), Jacobs et al. (1990), Saul and Van Ham et al. (2008) and Arya et al. (2004), who reproduced the 'primitive' in their visual representation of the Nagas, not only through revisiting their material culture and comparing them with Haimendorf's collection, but also by placing their ecology and livelihood within the framework of the 'ecologically noble savage' (Haymes 2007). These representations and caricatures of the Nagas as *jhum* cultivators and possessors of indigenous knowledge have found widespread appeal among NEPED department staff who have published literature and deployed the stereotypes of colonial imagination through sketch illustrations to showcase *jhummias*' local knowledge as valuable resources to build upon in community-development projects. However, in this engagement, the power relationship between the project staff and beneficiaries, the photographer and the photographed subject, the colonizer and the colonized, the anthropologist and his subjects of enquiry, has hardly changed. The optics and tropes of representation have remained quite the same as in colonial times. My research tried to unsettle such relationships between the fieldworker and the field. I have used history (written and oral) and narrative to shape a diachronic perspective of agrarian change, focusing on various actors, including *jhum* villagers, state departments, aid workers and project beneficiaries and institutions like the church, village bodies and the state. At the same time, I have contextualized Naga agriculture within the wider world of development practice that is shaped by the received wisdom of colonial knowledge and farmers' own understanding of continuity and change.

The book locates its analysis within a historical timeline beginning with the 1920s, when the Yimchungers and their neighbours were partially conquered by the British frontier administration, which came into contact with them during anti-headhunting raids. The Yimchungers became collaborators in a colonial pacification campaign directed against neighbouring villages that still raided and took heads and slaves for trade from the administered territory. The colonial phase was marked by a 'civilizing the savage' mission executed through pacification tours as well as bringing friendly villages within the political control and influence of the district administration. This established a new regime of control, as the colonial administration advanced its influence over the unadministered Nagas. However, more important than pacification was the nature of the relationship that built the knowledge of the Nagas and their relationship with the post-independence state, one that is marked even today as 'chaos' and 'statelessness'. Colonial contact during annual tours produced an image of 'savages' who were exposed to modernity through administration and missionization. The colonial administrators, who worked as administrator–anthropologists, played their role by documenting the Nagas' material and cultural artefacts. It not only primitivized the Nagas but also portrayed them as people of 'another time' – as a savage and backward race. The changes that came to Naga life and culture after the transfer of power to the post-independence Indian frontier administration were more remarkable and profound, as the process of territorializing frontier space and advancing state control through a militarized counter-insurgency programme was consolidated, in which the Nagas acted as collaborators as well as resisted state control through militancy and through negotiation for self-determination. With time, Naga nationalism has metamorphosed into ethno-nationalism. The genesis of ethno-nationalism can be traced back to the colonial project of ethnogenesis – ethnic involution through which the Nagas, like other hill tribes, were produced as the 'ethnic other'.

The political context of pacification and raids did not allow the normal administration to take over the Naga Hills. In this context, control was exerted through contact and the influence of local hill administrations. The deputy commissioners of the Naga Hills took enormous care to have personal influence over chiefs and village heads to control raids and settle disputes. The colonial administration in this process developed a vast sociology of knowledge through 'tribal monographs'. These documents became treatises for future administration. Thus, colonial knowledge of the Nagas was advanced through the anthropological writing of their customs, social life and culture.

Regarding *jhumland* control, the policies followed in the rest of the administered territories of British India could not be pursued here. To control *jhum,* the colonial administrators followed a new diplomacy of negotiation that was based on effective administration and recognizing the rights of the local people in land use and its management. I call it the 'science of empire', following Gilmartin's (1994) use of the term, or the logic of administration that was based on local customs and tradition through which colonial administrators took de facto ownership over land and legitimized its control and guardianship in the hands of village heads and chiefs. Major legislation like the Chin Hill Regulation III of 1893 and the Nagaland *Jhum* Land Regulation of 1946 were adopted to control *jhumming* following local sensibilities of place and culture. The principles of these legislations were further created under the constitution's Article 371 (a) based on the 16-point agreement between the Government of India and the Naga National Council that gave birth to the Nagaland state within the Indian federation. The special provision created tribal councils and recognized the village heads as custodians of the land, through which the state could reserve forestland only with the approval of consenting village heads who became go-betweens in such transactions, unlike in the rest of India, where the district administration and the Forest Department were, all-in-all, the owners of state forest.

In the Naga Hills, this difference created a new kind of brokerage between customary heads and the state that exerted its control over land through customary codes and conducts interpreted by government appointed *dobashis.* The *dobashis* or political go-betweens acted closely with the deputy commissioner and his lower ranks of administration to settle all kinds of disputes, starting from theft and robbery that involved two villages, to more significant village boundaries and land disputes (see Map 8.1). The customary codes of conduct present a distinct environmental history that employed local experts to train *jhum* cultivators in permanent wet-terrace rice farming and valued the use of alder pollarding in *jhum* management, unlike those followed in the rest of British India, where the plough was introduced to replace the hoe and *jhum* settlers were forced to take up *taungya* agroforestry. In the unadministered parts of the Naga Hills the *jhum* lands were kept outside the remit of taxation and, wherever the house tax or 'hoe tax' was imposed, it was not to exploit Nagas but was a measure to bring improvement and to promote sedentarization. The ideas of swidden being wasteful, unsustainable and soil degrading, picked up in the 1930s to 1940s as the *jhumland* legislation was passed, but control over *jhummias* were best kept in the hands of village headmen, who acted as intermediaries in administration. This process continued into the post-independence Indian state.

Interestingly, along with ideas of waste in the post-independence period, *jhum* was also understood as a practice that was unsuitable in light of growing pressure of population on the land, which could only be overcome through introducing new developments in farming. This has found resonance in the policy literature on agro-ecological development that links warfare and insurgency to the simple identification of forest people and *jhummias.* Various literatures produced by NGOs, government reports and academic writings claim the 'durable disorder' of the Naga Hills was retarding its progress and acting as a resource curse (e.g. the DONER-prepared *Vision 20:20 Report Northeast* also carries this view). *Jhumming* is inevitably blamed for statelessness and chaos. The FAO's *State of the World's Forests 2005,* observes that forested areas in poor countries are prone to violent conflicts and war. Forests

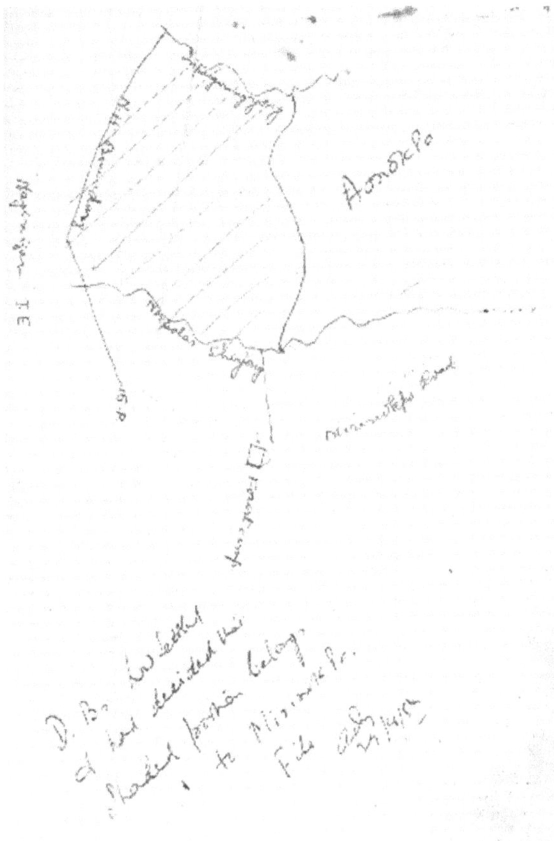

Map 8.1 Mokokchung villages land dispute maps with signatures of *dobashis*.
Source: Deputy Commissioner Office Records, Case Files, Mokokching, 1929.

and swidden landscape, as it is said, offers a hiding place and a source of revenue for insurgents, hence the presence of forests can become a motive for engaging in war, or in forests easily turning into breeding grounds for violence (FAO 2006: xiii). As explained in the report's final chapter, 'Forest and War, Forest and Peace':

> Forest regions in poor countries usually have many of the characteristics associated with the locations where violent conflicts occur. They tend to be remote and inaccessible. They often have valuable timber, petroleum, land, ivory, diamonds, gold and other minerals, which insurgents can exploit or tax. Forest dwellers frequently resent that outsiders benefit more from these. Governments have tended to see forested regions

> as peripheral places with few people and little political importance or economic value, and have only focussed on them to extract timber and minerals. Thus, forested regions have traditionally been poorly integrated into national political processes and receive few public services. Dominant ethnic groups have marginalized indigenous and tribal people in frontier regions who also compete with migrants from other areas over resources. Given the limited employment opportunity in many forested regions, taking up arms may seem like an attractive way to earn a living. (FAO 2006: 117)

This statement quoted clearly presents a picture of what contemporary *jhummias* face in Nagaland, a war that has entered ceasefire but still resembles and appears in everyday narratives of the local people. But such narratives provide little scope to understand why violence persists in a resource-rich region. I begin the thesis with my fieldwork encounter in Chapter 2, where I was engulfed in a violent context and later, as I went back to the study village, heard narratives from villagers of violent encounters with armed forces and underground militants as terror stories. This helped me to reflect back on the social context and political terrain of Naga–state relations since colonial times and how violence is embedded in local histories of colonization and social relations that shape local processes and agencies. I thus contextualize violence within the discourse of marginalization and development in which the Nagas are trapped by the politics of territorialization and nationalization carried out through electoral policies and development grants and aids in the name of development. In looking more closely at the FAO report, it becomes clear that it is not primarily an environmental condition that generates violence, but rather the political predicament of marginalized people in a resource-rich periphery who are challenged to transform their way of life.

Since the 1950s, the post-independence NEFA administration, besides engaging in their regular counter-insurgency operations, also monitored new development projects, such as the opening of new roads, and more importantly engaged the agricultural department and local administrators to promote the cultivation of rice, thus intensifying the incentives for settled agriculture. However, these projects rarely succeeded in bringing a major shift in people's livelihoods or cultivation choices until decades later, when missionary evangelism, the ideas of progress and moral values were attached to a new regime of work and toil and of notions of civilization among the Yimchungers.

I am thus prompted to ask: Why did the shift to rice farming and changes in food choices happen so recently? I explained that agrarian change was triggered by a set of policies and projects that came to take shape in the late 1970s, as neighbouring Sema and Ao missionary evangelists introduced new values and attitudes towards consumption. The changes were not sudden but became meaningful once upland improvement started with government seed supplies and agroforestry promoted through its Rural Development and Agriculture departments in the 1990s. Land came under permanent tenure in the *jhum* plots and provided added incentives to grow rice in small plots of land even by borrowing from kin, *khel* and clan members. The process was simultaneously supported by long-bean cultivation as a low commoditized cash crop. These changes can be traced back to upland (*jhum*) agriculture and the shifting resource regime in which access and control over land were determined by tenure arrangements that were becoming permanent. These interventions (terrace rice farming alongside plantation farming) tell that usufruct rights (such as the right to collect firewood) were now restricted as new plantations and horticultural projects were colonizing the *jhum* plots and extending, in the eyes of the project officials, a long and sustainable fallow management. The government's initiatives to promote cash crops in the 1990s, a shift in policy from wet-rice intensification in the 1970s and 1980s through subsidies, agricultural demonstration, seed supplies and incentives for irrigation and flood-control measures, helped in land consolidation once trees were planted in previously *jhummed* plots, and orchards were developed for oranges. The relay cropping of maize with long beans in many village plots sustained a long cropping period extending more than five years, thus lengthening the period of tenure. During my fieldwork year 2008–9, farmers reported that they were continuing with long beans on many farms, and that there was no need to keep the land fallow, even after ten years. This had forced farmers who could earlier easily borrow land in the *jhum* fallow to work as wage earners in the fields of their kin and *khel* leaders. For these farmers, getting a share in the lowland and terrace rice plots in return for household labour contribution to the owners' field during the sowing and harvesting season was the means through which households could now pay tithes to the church. People who could contribute maximum rice to the church received rewards from the pastor and became important functionaries within the church committees. The popularization of rice failed in the 1950s and 1970s, because labour was scarce for working in the wet fields. Moreover, wet-terrace rice had little value in the 'Feast of Merit' and rituals. Rice was one of the many crops grown in the first-year *jhum* fields but was not central to cultivation. The prestige attached to rice consumption 'as a crop of civilization', and one that connected the Yimchunger to the outside world, encouraged villagers to grow wet rice even in small plots. Rice consumption and farming have become symbols of people's improvement. My host claimed in his interviews that he and his brother were the largest paddy growers in the village and contributed maximum tins of rice to the village church, for which they were rewarded during the church annual competition on Christmas Eve.

The diversification of agriculture and the dependence on rice and *jhum* fields have led to rice intensification, although the returns remain low. The incentive to grow rice comes, as farmers explained, through its superior taste and through changing livelihoods in the village. Farmers can now opt for plantation agriculture, relinquish burning and even move to nearby towns and manage their fields by paying wages to village voluntary groups annually engaged by the village church during the sowing and harvesting season. The cultivation of rice has seen a phenomenal leap both in production and consumption. However, the production levels vary annually, as is revealed from church records due to the vagaries of weather (see Appendices 8.1 and 8.2). While, in 2009, monsoons were delayed, leading to poor harvests, in 2007 farmers remembered there were flash floods that damaged wet paddies and led to the loss of fish in the rice ponds. Despite annual variations, rice has grown in importance in peoples' diet along with long beans, which was not long ago a side dish, like fermented soya beans. Millet and Job's Tears are slowly becoming insignificant in farmers' diets, with declining consumption and a stigma attached to them as 'brewing crops'.

State farming policies and population pressure on the land have not been the singular cause of rice popularization and the intensification of agriculture. These relate to the changes in the resource regime from short tenure (two years under *jhum*) to long tenure under tree plantation and long-bean and maize-based relay cropping that have grown in favour of rice intensification based on the reciprocal exchange of labour in return for land. Population pressure on land, which is commonly understood to fragment landholding, is taking place in wet-paddy fields but not in *jhum* plots, because the tenure rights are different. In the wet-paddy plots, farmers divide their farms among their children, while in the *jhum* plots, land is inherited without fragmentation. Thus, population pressure was felt differently in *jhum* and paddy fields. Many of the landless families joined as wage labourers or were engaged as blacksmiths, porters in the towns and traditional basket makers in the villages. The villagers also migrated to neighbouring towns for work. I recorded in my household survey that at least one member of the village went to town for education, menial jobs and for other employment prospects. The village was no longer self-contained and insular from the outside world. The migration of people between village and towns due to marriage, education and job hunting was common. Population pressure on land was thus one factor in influencing agrarian change. Instead, the values attached to rice consumption and the moral economy of paying tithes to the church, both in cash and in kind, were producing enough labour to perform sowing and harvesting activities that were not possible with *khel* and household work parties before the coming of the church.

Simultaneously, the concentration of landholding in the hands of a few farmers was not leading to wholesale privatization of the land. Rather, as I have argued in Chapter 5, it has resulted in changes in swidden farming. It is one of my conclusions that enclosure of land through plantation and wet-terrace rice cultivation has not engendered absolute class formation; rather, it has led to a reconfiguration of the community based on multi-stranded personalized relations, where farmers and both landed and landless labourers are tied to one another through reciprocity and bonds of kinship and patronage.

In Chapter 5, I also explore how the 'commoditization' and 'privatization' of landholding are interlinked but do not take the shape of the wholesale individualization of landholding. Rather, I demonstrate that agricultural difference is predicated on the concentration of landholding by village elites, who manipulate access to land and the control over it, which is granted by new forms of territorialized land use (plantation farming amidst *jhum* fields and wet-terrace rice cropping co-evolving with swidden). The chapter centres on the role of go-betweens, by focusing on their life histories, and demonstrates how social relations and political patronage shape farming in the hills. Thus, *jhum* cultivation among the Yimchunger Nagas presents a diachronic history other than what is perceived by planners and agricultural experts. Over time, people adjust to the changes coming through land shortage by shifting and consolidating land and by planting new crops that are introduced and promoted by state agencies.

New technological interventions, the demand for cash, and the idea to be 'thrifty-economical-individualistic' promotes the privatization and individualization of landholdings through land consolidation, as it does with patterns of consumption and feast giving. During my stay in the village, Metümneo feast-giving had ceased and was regarded as a practice of the past. But feasts were still given by village headman on the birth of a new child and during the baptism of new-born babies by the invitation of a selected few who were close kin members, village headmen and go-betweens in village development work. This has resulted in the collapse of communal activities that involved the whole village in the past. Here, the static notion of *jhum* agriculture as 'primitive farming' has been markedly undermined by strategic labour choices made by the Yimchunger swiddeners (*jhum* families) that sustain and help co-evolve paddy farms alongside *jhum* plots as composite farming (Rambo 1998). The physical limits placed by the increasing pressure of population on land have historical and administrative connotations (the ways in which villages became fixed to a particular place) and are equally determined by changing land relations and values attached to land as property. These shifts further transform once-subsistence economies to more integrated market economies that make families and individuals central to the whole 'community'. Individualism, privatization and an economy of resources then become inevitable, leading to the accumulation and individualization of resources. Reciprocity and barter exchanges do exist between interest groups that foster patronage (e.g. the church, and state officials).

Contemporary studies on shifting cultivation in Nagaland, though very few, have produced the post-independence narrative of population pressure and the neo-Malthusian explanation of non-sustainable agriculture (Langkua 2004). Other scholars have found *jhum* innovation, based on age-old indigenous knowledge of *jhum* fallow management, to be a sustainable intensification practice that can be replicated all over Nagaland (Cairns et al., 2007 a,b, 2011). The NEPED project also began with the idea that *jhum* innovation developed by Angami Nagas and other Tynemia groups can be taken up as a local sustainable practice to intensify land use, promote conservation and effective fallow management that do not put pressure on the *jhum*-based agro-ecosystem. However, the NEPED project planners themselves broadened their concerns in the second phase (2001–6), where they incorporated the economic aspects of sustainable agriculture through the promotion of soft loans (revolving fund) and introduced the concept of self-help groups and revolving funds to build upon Naga *jhum*.

In my study village, I discovered one farmer practising alder-based *jhum* fallow management, which was the cornerstone of the NEPED project intervention in Phase I (1999–2001) all over Nagaland. My interaction and interviews with farmers revealed that they were not ignorant of the alder plant's unique qualities in soil conservation. However, they never introduced the crop in their fields. Planting these trees, as farmers explained, hindered their immediate returns from agriculture. As alder-based *jhum* fallow management requires a number of years to establish, farmers found it economically non-viable in the short term. The only farmer who took up alder-based plantation in the study village was self-motivated, and much hearsay and gossip made me believe that he was the sole beneficiary of the introduced alder plantation project in the village. His farm became the test plot and remains a single speck of well-managed sustainable *jhum* plot based on alder (see Image 8.1).

Malcolm Cairns et al. (2011) describes composite farming, based on their detailed field research in two Angami villages in Kohima district, as a historically evolved practice among the Angami Nagas. On the contrary, my research shows that a new type of 'composite swiddening' has emerged in Leangkonger village, where people have been introduced to wet-terrace rice farming since the late 1920s. However, villagers have started taking an interest in *jhum* innovation based on the dual cultivation of wet rice and *jhum* crops

Image 8.1 Alder-based plantation in Leangkonger village.
Source: Author's photograph, 2009.

(based on the relay of long beans and maize) quite recently, as these crops have become important in their consumption habits with an incentive to sell and earn cash. Here, the church, the neighbouring market demand for long beans, and villagers' aspirations to emulate their neighbours (advanced Ao and Sema Nagas) have led to rice intensification and consumption leading to an exchange of land and labour between *jhum* and wet-paddy fields. All this is underpinned by labour exchanges that are shaped by emerging resource regimes, where enclosures of upland farms for plantation push farmers to intensify agriculture through reciprocity of land and labour.

Thus, I infer that the *jhum* cycle is highly innovative and adaptive to change. The synchronic narrative of *jhum* under population pressure does not present an acceptable picture of the different kinds of fallow management that have changed over time, not only due to population pressure but also because of state policies and people changing their habits towards consumption. Also, there have been different practices of swidden management under traditional *jhum* that have been largely overlooked in policy debates on *jhum* improvement that have defined *jhummias* as forest destroyers and settled wet-paddy cultivators as forest conservers (see, e.g. the chapters on *jhum* management prepared in DONER Vision 20:20 for North East). To cite one small example, when Hutton made his journey to the Yimchunger village, Yonghong, he described the Yimchunger practising *jhum* much more intensively than they do today. On 18 April 1923, he writes:

> Indeed the *jhumming* system of the village round here is about the finest I have seen. Only millet and Job's Tears are grown, but the whole hillside, and very steep it is, is most elaborately laid out in ridges and quasi-terraces with logs cut from the pollarded alder growing all over the slopes and everywhere most carefully preserved. Unlike the *jhums* of other tribes, which are used for at least two successive years, the ground is sown for one year only, and, then allowed to stay fallow again for three or four years, instead of the more usual ten, and this rotation is continued with apparently admirable results, showing what really can be done with steep and unpromising land by careful preservation of the alder and precautions against denudation. (1924: 28–29)

The observation by Hutton clearly highlights how some villages could intensify agriculture through a more efficient fallow management using alder in a non-Angami area. Population pressure was not the guiding force behind such *jhum* innovation in the colonial unadministered Naga areas. Rather, it reflected the diversity of fallow management under *jhum* practised by different Naga tribes. Today, the landscape is slowly becoming homogenized through the introduction of plantation farming and agroforestry, which has been partially successful. Similarly, wet-terrace rice farming has been taken up in certain villages, while it was not so successful in past days. The reasons are varied and specific to village agroecology and farmer's needs, the common perception among Yimchunger Nagas is that they should consume rice, as it brings them closer to their neighbours and places them on the path to development.

As I have discussed in Chapter 5 as well as Chapter 6, rice has become the farmer's principal crop of consumption in Yimchunger villages, despite its secondary status until quite recently (unlike in Angami areas where rice has been the staple food crop of the people for generations). The popularization of rice as a food crop also comes through its value and as a prestige crop of civilization, improvement and a marker of development. In this process, the evangelical missions in the Yimchunger area by Sema and Ao neighbours have played a leading role in village improvement. The missionary's influence over the Yimchunger area was unique, as the missions were carried out by Nagas themselves and were done in the period just after India's independence. The mission work progressed among the Yimchungers with grassroots participation of the first converts, who were the village heads. When the original pastors went back, the vacuum left was filled by the village headmen, who served as pastors in the village church. In this way, the Baptist missionaries worked with the grassroots converts. The missionary's persuasion to grow rice was conditioned by the changing resource and labour regime in the village and their vision of work, time and toil. The missionary-organized wage groups participated in a government public works programme in the 1970s to earn wages for the construction of the village church. The village work force was also organized to motivate villagers to grow rice. Rice consumption has become popular in recent times, as rice has emerged as a prestige crop. People who contributed maximum rice to the church were rewarded to grow more. It was also used as a symbol of prosperity and modernity by villagers who contribution rice instead of cash during the church golden jubilee celebration in 2008. Past failures to popularize rice in farmers' cuisine were determined by both contributions of labour and people's attitude towards rice consumption. In the contemporary context, missionaries promote rice over millet. The change in farming practice thus needs to be looked at as the changes that have come in the Yimchunger Naga way of life over time, looking at the new world views introduced by the missionaries.

During my PhD fieldwork, I realized that farmers were sharing the received wisdom of soil loss and ecological decline as defined by 'experts' and 'development workers'. The first question I pose then, is how do farmers take such 'received wisdom'? There were many instances during my fieldwork when farmers challenged the 'experts' knowledge and relied on local knowledge. For example, farmers were reluctant to adopt the *Nepalese alder* as *jhum*-stabilizing trees introduced by environment NGOs and the state-donor run NEPED project, but were enthusiastic to experiment with cash crops. The commoditization I witnessed during my fieldwork is best described by the term 'low commoditization' (where crops are grown both for household consumption and as cash crops) as used by Bernstein and Woodhouse (2001). Low commoditization through long-bean cultivation as a relay crop with maize and millet has significantly changed cropping patterns and land use from millet, mixed crops and Job's Tears to long beans grown in combination with corn during a single-year crop cycle. This type of low commoditization of crops helps to generate both subsistence crops and diversify farmer's cash income in an increasingly monetized village economy. The introduction of highly commoditized crops in the recent past, such as ginger and cardamom exclusively grown for the market, failed in the absence of market demand, resulting in farmers' non-participation.

On the other hand, success as constructed by the project staff was also reflected in people's day-to-day discussions, where Leangkonger men claimed theirs to be the most successful village. In reality, the village was short on firewood and grain production compared to the neighbouring settlements that were remote and less connected with development work. Farmers explained that the village had limited forest area and cropland. But the problems were also accentuated by the changing resource use: they were engaged in producing surplus for the market and at the same time diversifying agriculture that would attract more and more government projects. The experimentation with many exotic crops, such as jatropha, ensured that there were no immediate returns and that land was sealed off from food crop production for many years by the village landed elites. Similar was the case with timber and horticulture plantation. Farmers who took the initiative to promote these crops were also brokers in government rural development programmes and were private Self Help Group organizers in the village. By introducing these crops, they became suppliers of seeds, subsidies and loans to *khel* members and legitimized their leadership as benefactors, regardless of their social position as second settlers within the village. My host's son's interviews make this point clear. Responding to my query on why his father was not successful in building a new *murung* house, he replied that

> village heads were the first settlers, owners of the village and reserved the sole right of building a *murung* house. But where did the money for building a new *murung* house come from? My father is a political party representative in the village and had the skills and privileges to work as a go-between in various development schemes that are aimed at village development. Village heads do not have such capacity to bring about development in our village, yet they claim to be our heads.[1]

This observation clearly reflects how village power relations are shaped and passed on to new actors who act as village intermediaries and claim patronage over the village, challenging the traditional authority of the village headmen. The introduction of new farming technology and crops gave opportunities to villagers who are now educated and participating in political activities in the village.

These programmes were not only changing the face of swidden agriculture but were also becoming instruments through which the villagers established their positions among clan members. During my stay in the field for 14 months, I travelled widely between villages, towns and district headquarters in Mokokchung, Kohima, Mon and Tuensang, and in each of these places I identified a diversity of change in farming away from what is conventionally presented as 'slash and burn', a pejorative term that did not match my understanding of Naga agriculture.

Burning is an integral component of *jhum* but not the only one. Even in Kohima (Angami Naga villages), where farmers have excelled in alder plantation-based fallow management, forest fires would make headlines in local newspapers during the pre-sowing season. *Jhum* cultivation is indiscriminately attached to the worries of changes in temperature, soil erosion and floods. These have become standard ecological narratives. Even the NEPED team, which made visits to Shamator town in a new initiative of biodiversity conservation with the SACON fieldworkers during 2009, were likening climate change events to the 2009 late arrival of monsoons that delayed the sowing of wet rice in many Yimchunger villages and led to crop damage. These uncertainties and weather events were now blamed on deforestation caused by *jhum* and burning. These narratives also catch the imagination of *jhummias,* who see *jhum* as less productive than the incentives that come with tree plantation and wet-rice farming in the form of subsidies, credit assistance and new seed supplies (see Image 8.2).

This has led to various kinds of innovations in *jhum* plots. In Leangkonger innovation has taken the shape of composite swiddening, where both wet-terrace and *jhum* crops are grown annually. In other parts of Nagaland, such as in the Ungma and Kupza villages of Mokokchung district, which I visited during fieldwork, quarrying, pig farming and the plantation of tea, pineapples and the cultivation of vegetables such as tomatoes, have integrated the village with the local market and district headquarters. Farmers are more dependent on the export of tea and pineapples for processing to local canning factories set up in Dimapur. *Jhum* fallow has become permanent, and the *jhum* cycle has reverted to a healthy 20-year cycle, as one agriculture department brochure notes. *Jhum* innovation in Khonoma has been well documented in studies carried out by Cairns et al. (2011), Cairns (2007a,b) and NEPED (1999; 2007). They have identified it as composite farming now under pressure from migration in villages close to towns and urban centres. My observations also reflect how composite swiddening has evolved in Leangkonger. However, the impact of such farming on local agronomy and households was impacting people's accessibility to land and resources that were earlier not territorialized by villagers who have the resources and command to bring them under permanent cultivation.

The study thus explores new grounds in the social interface of ecological change. It throws light on environmental change by presenting alternative reading on shifting cultivation from historical, oral, temporal and visuals (photographs) of the colonial times and ethnographic reflection from the present. It is my conclusion that the linear-evolutionary history of agroecology is deficient in many ways for understanding the dialects of agrarian change. Thus, we need to challenge the received wisdom by paying particular attention to farmers' tacit understanding of the environment, the moral economy of households, the micro-politics of resource use and local histories of agrarian

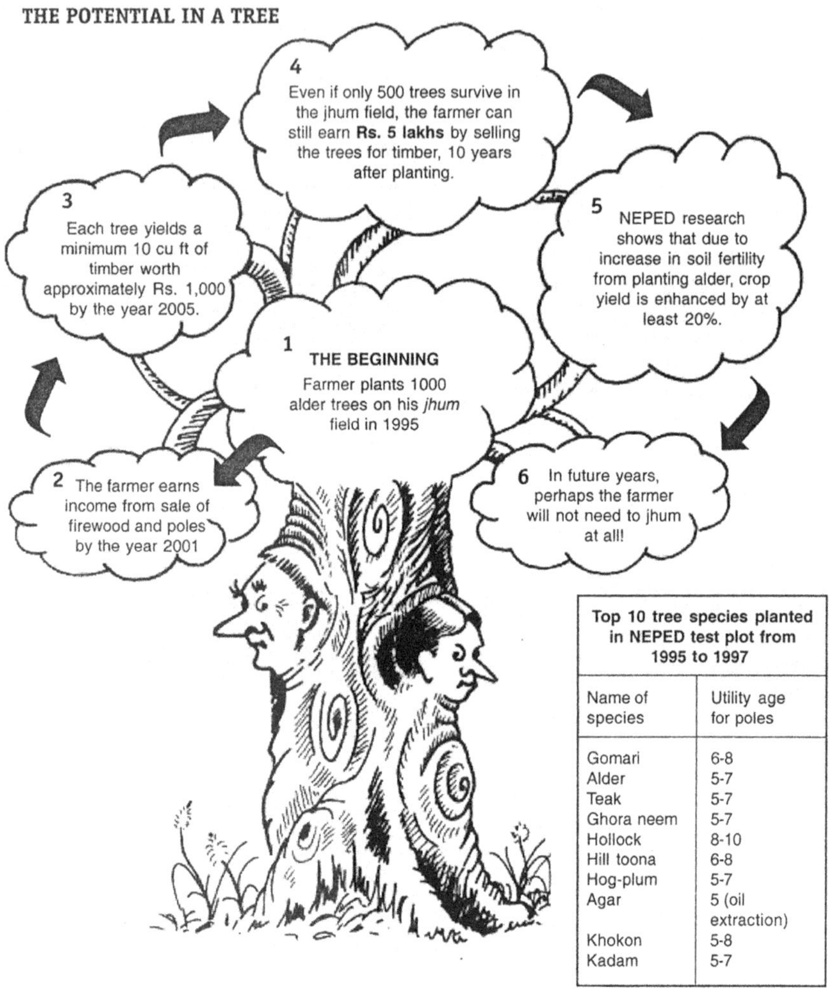

Image 8.2 A poster depicting potential of a tree.
Source: NEPED, 1999.

change. This study, thus, looked at agro-ecological change in a Naga village by taking a diachronic perspective on agrarian development in the Naga Hills in general and the Yimchunger study community in particular. My research has persevered to bridge the gap in the contemporary literature on highland swidden agriculture that has looked at the discourses and narratives of a dominant policy paradigm floated by transnational environmental NGOs and institutions, but does not pay attention to the micro-policies of the swidden farmer's life worlds. The results from my analysis may help planners and project implementers to design policies and programmes on swidden improvement and help fill the gap in studies of agrarian change in highland Nagaland and Northeast India. I have presented both the indigenous perceptions of *jhum* and how the state and experts frame it as a particular mode of subsistence centring on its inefficacy and underproductivity. This ethnographic description of the changing social life under *jhum* agronomy will help to redress such negative stereotypes and engage with *jhum* innovations while designing holistic agro-ecological policies and programmes on *jhum* land development.

Appendix 8.1 Production of crops in the study village (unit used in production is tins, 1 tin equals 15 kgs)

Crops	1991	1992	1993	1994	1995	1996	1997	1998	1999	2000	2001	2002	2003	2004	2005	2006	2007	2008	2009
Paddy	765	768	720	666	980	777	978	777	888	1200	780	885	999	1500	1450	1300	1550	1600	1000
Millet	900	499	550	127	356	980	346	350	459	551	111	367	300	600	450	380	380	400	250
Maize	709	328	776	256	980	666	439	400	200	309	84	147	265	400	260	150	260	250	300
Job tears	450	576	470	796	330	470	306	576	300	150	53	68	60	100	80	60	45	60	50
Kholar	199	265	360	400	888	380	236	486	200	444	220	444	220	200	150	180	160	270	100

Source: Leangkangru Baptist Church Records kept in the village on annual contributions to the church.

Appendix 8.2 Production of all crops.

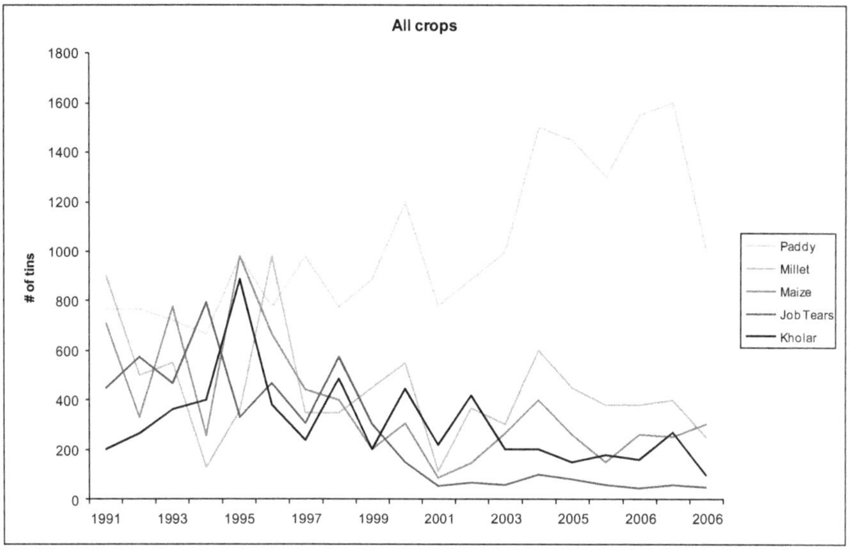

Source: Leangkunger Church records.

Appendix 8.3 Recognised Naga community as scheduled tribes, 1991 census.

Nagaland	Manipur	Arunachal Pradesh and Assam
1. Angami	1. Anal	1. Tangsa
2. Ao	2. Chiru	2. Wancho
3. Chekesang	3. Chothe	3. Nocte
4. Chang	4. Kabui (including Puimei)	4. Zeliangrong (Assam)
5. Chirr	5. Zema and Liamgmei	
6. Khiemungan	6. Koireng	
7. Konyak	7. LamgangMao	
8. Lotha	8. Mao	
9. Makhori	9. Maram	
10. Phom	10. Maring	
11. Rengma	11. Monsang	
12. Sangtham	12. Moyon	
13. Sema	13.Tangkul	
14. Tikhir		
15. Yimchunger		
16. Zeliang		
17. Pochury		

Note: According to other sources the number of Naga tribes vary. Yonuo says there are about fifty 'tribe' (1974:6–7). Atai Shimray lists 40 Naga groups (Shimray 2005:32–33). Horam says, "there are as many as 32 known Naga tribes" (1975:27). The Naga Ho Ho claims that there are 36 Naga tribal groups.

Source: Gazetteer of Nagaland, Manipur, Arunachal Pradesh 1991.

NOTES

1 Introduction

1 Georges Condominas, 'Anthropological Reflection on Swidden Change in Southeast Asia'. *Human Ecology*, 37 (2009): 265–67.
2 The term 'Zomia' was originally coined by Van Schendel (2002) and adopted by James C. Scott (2009). According to James Scott, Zomia is a new name for virtually all the land at altitude above roughly 300 metres, all the way from the central highlands of Vietnam to Northeast India and across many Southeast Asian nations (Vietnam, Laos, Thailand and Burma) and provinces of China (Yunnan, Guizhou, Guangxi and parts of Sichuan). It is an expanse of 2.5 million square kilometres containing about one hundred million minority people of truly bewildering ethnic and linguistic variety. Geographically, it is also known as the Southeast Asian mainland massif but, politically, the term refers to the fact that these areas have been beyond the control of governments – unlike lowland areas.
3 Christoph von Fürer-Haimendorf archival notes, School of Oriental and African Studies, University of London. PPMS, Box 31, 11–12.
4 See note by C. U. Aitchinson, secretary to the Government of India, Foreign Department, in 'Demarcation of the Duffla Frontier; Foreign Department, Political A,' April 1874, Nos 268–69, NAI (National Archives of India).
5 I use missionary modernization in a particular sense to describe the changes that have come into Naga people's lives though changes in their food habits, world views, consumption of rice and their understanding of time as organized by the church calendar. This has added new meaning to peoples' understanding of leisure and toil. This is discussed in detail in Chapter 6.

2 Methodology and Fieldwork: Negotiating Hazardous Fields

1 During my fieldwork there were three major political underground factions operating in the area. Two factions of the National Socialist Council of Nagalim: (1) a faction led by Thuingaleh Muivah and Isaak Chisi Swu, the National Socialist Council of Nagalim (Isak–Muivah) (NSCN–IM); (2) a different faction, the National Socialist Council of Nagalim–Khaplang (NSCN–K), and (3) the Federal Naga Army (FGN), as well as many other groups, such as the Naga National Council (NNC) and also self-styled individual claimants who occasionally appeared in the village claiming allegiance

to one of these groups. The Government of India has separate ceasefire agreements with each of the main groups, the first having been made in 1997 with the NSCN–IM. These agreements have been renewed periodically.

2 Philip Bourgois, while doing his fieldwork in El Salvador, became partisan with his research community to an extent that he started publishing in American print media the fieldwork encounter of counter insurgency, reporting from the field. On returning to the university from fieldwork, he was fired by his department, accused on many counts of having breached anthropological research ethics and the code of conduct set by AAA. Bourgois writes: 'Significantly, had I not gone to the media with my testimony of human rights violation, would not anthropological ethics have been violated in a serious manner? It would have remained a personal story between my committee and myself as an unsuccessful and reckless preliminary fieldwork exploration that had been decided against as too dangerous. By remaining silent I would not have violated anyone's rights to privacy nor have threatened my colleagues access to the field, nor offended my host country government' (Bourgois 2007, 295).

3 The Baptist and the Catholic *khels* lie some 2.5 kilometres apart. The Baptist *khels* form the main village while the Catholic *khel* is located away from the main settlement close to Shamator town.

4 Interview with Anumgla in Khonoma village, February 2009.

5 A classic example of this kind of insensitive, journalistic, writing is D. R. Mankekar's *On the Slippery Slope in Nagaland* (1967) and a recent one by Jonathan Glancey, *Nagaland: A Journey to India's Forgotten Frontier* (2011). There are dozens of other books that have misrepresented Naga aspirations and the resistance movement and made life difficult for people who are genuinely committed to fieldwork and telling the story of the Naga resistance movement and the cultural politics of violence in Nagaland.

6 J. H. Hutton, who in 1923 conducted his Trans Dikou tour into the territory of Kalyo-Kengyo (Kheimungen) and other Naga tribes in the unadministered territory, travelling with his botanist friend, Kingdon-Ward, recording the details of his tours and his visit to Yimchunger villages. The narratives of two of these tours were subsequently published as *Diaries of Two Tours in the Unadministered Area East of the Naga Hills* (1929). Following Hutton, J. P. Mills, the deputy commissioner of Naga Hills District and a prolific writer on the Nagas, conducted a punitive expedition in 1936, accompanied by Fürer-Haimendorf.

7 In 2007, an Ao Naga PhD student from Nagaland University had to provide a feast for his study village respondents for misrepresenting their culture. His writing cost him both intellectually and financially. This incident was narrated to me by his friend, a local lecturer in Kohima College. It is quite unfortunate that I could never meet this research scholar to get firsthand information of his fieldwork, as he passed away while I was doing my fieldwork.

8 Nagaland is part of a global biodiversity hotspot as well as an eastern Himalaya endemic bird area. Bird-watching and ecotourism associated with it has become very popular in Khonoma village and Fakim Wildlife Sanctuary. Tour operators from the plains of central India liaise with local villagers and travel agents in Kohima and Dimapur to organize such trips. During my fieldwork stay in Khonoma village I was an eyewitness to two bird-watching groups who came from Europe with a local tour operator and one from Goa.

9 *The Illustrated London News,* a mid-Victorian era journal, provided readers with rare images of the Nagas along with other tribes of NEFA and central India, among whom

Fürer-Haimendorf conducted his fieldwork. See PPMS 19, Box. 42. SOAS Archives. Fürer-Haimendorf collection.

3 Ethnography, Violence and Memory: Telling Violence in the Naga Hills

1 Access John Henry Hutton's biography at the Pitts Rivers Museum history link: *prm. ox.ac.uk/collector hutton.html.*

2 The hymn was sung, translated and transcribed by Sanghpur village elders, mostly members of the village council. Sixteen elderly men participated in singing and narrating the events that unfolded before and after the battle. Interview 27 July 2009, Sanghpur Village, Nagaland.

3 The people who lost most of their family members in the British raid later migrated to other villages such as Leangkonger, which was established in 1938. One of the lineages from the *Jangrü* clan that numerically dominates the Leangkonger population comprises descendants from a Sanghpur migrant.

4 Butler, the Governor of Burma at the end of the Burma campaign, earned the title of 'Great White Chief' for his leading role in pacifying slavery in the Shan, Kachin and Naga unadministered territories.

5 Harcourt Butler's Burma campaign was well documented by the mid-Victorian press. See, British Library collection IOR NEG 11049 and Mss Eur F 116/84 Butlers Collection newspaper cutting on the abolition of slavery and human sacrifice in the Hawkang Valley, etc. 1925–28.

6 The Treaty of Yandaboo in 1826 that led to the defeat of Burmese Kingdom brought Assam under the direct control of the British East India Company and then the British Indian Government after 1857. The British interest in finding a route to Manipur via Naga Hills brought them into direct confrontation with the Angami Naga tribes of the North Eastern frontier for the first time in 1840.

7 The Naga Hills District initially included the present Mikir Hills, western Rengma, the North Cachar Hills and part of the present Nagaland. Between 1875 and 1923 it became smaller, with territories transferred to the North Cachar Hills, Nowgong and Sibsagar districts (Shimray 2007: 232).

8 The regulation also set the territorial limits of revenue administration of the British Indian Empire in the foothills and 'exempted concerned bureaucrats from liability for non-application of the Indian Penal Code' (Chatterjee 2007). Over the period 1873–1930, the 'Inner Line' was modified several times to include valuable forest under reserve, mines and tea plantations and to create space for the Assam Bengal Railway in Dimapur *mouza.*

9 See, IOR NEG 110047. IOR Mss Eur F 116/82. Butler collection, bound volumes entitled 'Abolition of Slavery' containing copies of correspondence, memoranda, notes and reports on the expedition and IOR NEG 11048. To abolishe slavery and human sacrifice in the Hawkaing Valley, the Naga Hill Track and the 'Triangle' of Sir Harcourt Butler, J. T. O. Barnard, H. J. Mitchell and J. P. Dewar, 1924–27. IOR Mss Eur F 116/83. Butler Collection. Report and photographs of Sir Harcourt Butler's expedition to the Huwakang Valley and proposition to Abolish Slavery and Human Sacrifice there; J. T. O Barnard's 'Report on the Expedition to the "Triangle" for the liberation of Slaves', with photographs by P. M. R Leonard, a member of the expedition. J. P. Dewar's Report on the Naga Hills (Burma) expedition for the abolition of Human Sacrifice (1925–27).

10 The records of British officials rewarding Naga headhunters for taking Japanese soldiers' heads can be found in correspondence dated 14 July 1944. Ref. WO 203/4652/ Brigadier General Staff, 11, Army Group, SEA, 14 July 1944 to HW 14 Army. Public Records Office, Kew Botanical Gardens, London. This military strategy was also followed in Malaysia, where the British frontier administration was losing control over highland people. Tom Harrison, a British anthropologist was deployed in the frontier to counter the Japanese invasion and he encouraged and trained Kalibet highlanders to take heads.

11 Fortnightly report in the Assam Tribal Area for the second half of September, p. 139. As appeared in NEFA. Fortnightly report on the Assam Frontier Areas. IOR L/PRS/12/3117.

12 See Ministry of External Affairs Letter No. F-64/1/59-NU, dated 25 February 1959.

13 The Village Defense Guards were initially known as Village Scouts in 1957 and later in the 1970s were renamed as the Village Guards.

14 As reprinted in EX-CBI Roundup, June 1956, 'India's North Eastern Frontier', pp. 8 and 10.

15 Man Bahadur Rai was the assistant political officer of NEFA's Tuensang Frontier Division and was a key figure in mobilizing the first batch of Village Scouts and the Assam Rifles. In his tour report of 1956–57, Rai particularly describes these areas as 'out-of-the-way' places like his colonial predecessors.

16 IOR L/PS12/ 3116/ North Eastern Frontier – Burma–Assam Border expansion of Control Area of Naga Hills 1947, p. 25.

17 Tour Report of Shri Man Bahadur Rai, Assistant Political Officer, Noklak 1956. Government of India, North Eastern Frontier Agency, Political Department-Confidential Branch, p. 42.

18 Ibid., p. 18.

19 Ibid., p. 20.

20 Ibid., p. 42.

21 Ibid., p. 55.

22 See correspondence dated 14 July 1944. Ref. WO 203/4652/ Brigadier General Staff, 11, Army Group, SEA, 14 July 1944 to HW 14 Army, PRO, Kew Botanical Gardens Records.

23 See, Confidential Department Branch, Regrouping of Villages in the Naga Hills, 1957. File No. c.384/57, pp. 4–5.

24 Tour Report of Man Bahadur Rai, APO, Noklak Tuensang Frontier Division, 1957. Political Branch, Confidential, File No. Tn/R. Arunachal State Archives, p. 24.

25 Ibid., p. 24.

26 See letter No. 130000/25/GS (O), dated 27 June 1957 from HQ GOC Assam company to Commissioner, Naga Hills Division, Kohima, p. 15. As seen in 'Regrouping of Villages in Naga Hills'. File no. C. 384/57. Confidential Branch, Secret, Department of Art and Culture, Record Cell.

4 *Jhum* and the 'Science of Empire': Ecological Discourse, Ethnographic Knowledge and Colonial Mediation

1 The village development boards were first set up in the 1980s. Subsequent legislation enacted in 2002, the Nagaland Communitization of Public Institutions and Services Act 2002 passed by the state legislature, made way for the neo-liberal doctrine of 'government through community'. This has facilitated financial decentralization and

made communities responsible for essential services like primary education, child care, water supply and sanitation and electricity distribution and its management in villages.

2 References to crop production frequently come up in the Annual Administrative Reports of Mokokchung and Kohima divisions of Naga Hills district from the late 1920s.

3 The Assam Forest Regulation 1891 lays down the rules under which local people could practise *jhum* within forest areas, to be determined on a case by case basis by the Forest Settlement Officer. It is noted that this 'is subject to control, restriction and abolition by the State Government' and was not a right. http://www.sfri.org/images/general/article_law_gns_slub.pdf.

4 See Letter No. 938, dated 12 June 1885 from the chief commissioner of Assam to the secretary to the Government of India, Foreign Department, pp. 89–91. As seen in Correspondences Regarding Issue of Inner Line Regulation Between Southern District of Seebsaugor and the Naga Hills, Record Cell, Nagaland State Secretariat Archives, p. 6.

5 The forest villages were formed by an immigrant labour force that was brought in for 'working plans' and 'taungya forestry'. Evidence of the former is more common in the Naga Hills that led to disputes between the new settlers who came in conflict with the Naga inhabitants over rights to cultivation. Mikoko Kumara (2008), who worked on land alienation among forest communities, cites many cases of land lost by the native tribes due to colonial policies that alienated tribal land.

6 From the officiating political agent, Naga Hills to secretary to the chief commissioner of Assam, File No. 103, dated 30 May, Samogooditing. Annual Administrative Report of the Naga Hills, 1873–74, p. 1. Department of Art and Culture, Nagaland State Archives.

7 The forest reserves included Dosai Valley, Nambor, Rengma, Diphu, Dhansiri, Tiru Hill, Geleki, Disai Reserve, Rangapahar, Itangki Reserve. To see a list of the reserves, see Ramunny (1988).

8 The villages in Dimapur mauza that were settled through *patta* (land deeds) were classified into Bari, Taklabari, Sali, Lahi and Faringati. These classifications of land categories were further placed as 1st class, 2nd class and 3rd class land. The assessments for each of these classes were factored differently at different rates. In Dimapur mauza by 1939, 4,500 *bighas* (unit of land measurement in Assam) has been settled out of the 29,000 and odd *bighas* that were primarily 'jungle' wastelands. There were 16 villages grouped in Bari land, 2 in Taklabari, 3 in Sali, 4 in Lahi, 5 in Faringati. See Government Memo no. 1801, dated 19 September 1939, Shillong, Revenue Department, Excluded B files. As seen in record no. 754–61, p. 6. Assam State Archives, Guwahati.

9 According to settlement and revenue officials for the financial year 1938–39, the household assessment of revenue accounted from *non-pattadars* and labourers, etc., was Rs 1350 and from *patta* holders was Rs 647. The *jhum* cultivators definitely outnumbered the 'revenue' (*patta*) landholders. The colonial officials were making a case for raising the revenue rates for patta land, as this would significantly increase the revenue base from the mauza. According to revenue and land resettlement officials, the revised rate would increase the land revenue from Rs 674 to Rs 1,734, which would further surpass the non-revenue (house tax) land assessment. This was important in light of the Assam province deficit revenue of 10 lakh rupees that financial year as much money had been spent on flood-control measures. See Government

Memo no. 1801, dated 19 September 1939, Shillong, Revenue Department, Excluded B files. As seen in record no. 754–61, p. 6. Assam State Archives, Guwahati.

10 See Government Memo no. 1801, dated 19 September 1939, Shillong, Revenue Department, Excluded. B files. As seen in record no. 754–61. Assam State Archives, Guwahati.

11 'In Nagaland, the notified reserve forests cover only 3% of the forests. About 93% of the forests are unclassified of following categories: (1) private land, (2) clan land (3) morung land and (4) common village land. The original inhabitants have absolute right to practice shifting cultivation and grazing their cattle over recognised shifting cultivation areas under the Naga Jhum Land Regulation Act, 1946.' Report of the Working Group on Forests for the Environment and Forests Sector, GoI-Planning Commission 2006: 8.

12 Colonial science, as Ravi Rajan (1998), Dipak Kumar (2006) and Richard Grove (1995) have argued, produced a body of knowledge that was used to promote the idea of conservation and resource exploitation in the colonies. By 'colonial science' here I mean particularly 'science' that developed a corpus of knowledge fashioned to support colonial situation and essentially came 'from the centre'. Colonial scientific knowledge was used for 'regenerative forestry', 'prospecting minerals' and other resources that could help build the empire and sustain its extractive industry in the colonies.

13 As noted earlier in this chapter, a substantial portion of the virgin forest in the foothills near Dimapur, Amguri, Simulguri, Golaghat and in Sibsagar district had been declared reserve forest.

14 Three kinds of official sources document the official stance towards *jhum*: The Annual Reports of the Assam Forest Department, the Annual Administrative Report of the Naga Hills published during the late nineteenth century, tour diaries of annual visits to British-administered villages that continued to be written well up to the 1980s contain many negative descriptions of jhum and contrasting positive appreciation of paddy farming, as it did not disturb soil and forest during cultivation.

15 Gilmartin's (1994) imperial science/science of empire contrast deploys the same central/local and high/low categories as the debate described above about competing discourses within the natural sciences – which makes it somewhat difficult to relate the two debates to each other.

16 See the *Illustrated London News*, 1937.

17 The image of the ecologically noble savage has been strengthened by the NEPED projects promotion of *Napalese alder*-based fallow management that has been celebrated since colonial times as the most innovative way of developing *jhum* from within. The innovation of alder-based fallow management is also reflected in Cairns (2007a, 2007b). Cairns and Brookfield's work (2011) claims Angami swidden practice as the most sustainable indigenous fallow management practiced in Nagaland.

18 Annual administrative report of the Naga Hills, 1869, no. 1 and 2, p. 11, Record Room, Department of Art and Culture, Nagaland Government.

19 Annual Administrative Report of the Naga Hills District 1884–85, p. 7, Record Room, Department of Art and Culture, Nagaland Government.

20 Annual Administrative Report of the Naga Hills 1884–85, p. 3. Record Room, Department of Art and Culture, Nagaland Government. (In 1884–85, the Naga Hills

was incorporated in the Assam valley districts. The Annual Administrative Reports were now directly included in the *Assam Gazetteer* records. However, these administrative changes did not significantly alter the nature of revenue administration.)

21 Letter from H. S. Beadon, Esq., Offg. Under Secretary to the Government of Bengal to the officiating Commissioner of Assam, dated 17 February 1871, p. 4. As seen in the Government of Bengal, General Department File, titled Potato Seeds for Naga Hills, File No 206/364 of 1871, Record Room, Department of Art and Culture, Nagaland Government.

22 From Lieutenant John Butler, officiating Deputy Commissioner of the Naga Hills, to the Personal Assistant to the Commissioner of Assam, dated 10 October 1871, as seen in file titled 'Potato Seeds for Naga Hills', File no. 206/364, Record Cell, Department of Art and Culture, Government of Nagaland.

23 Ibid., p. 100.

24 File no. 275/981 of 1923 office of the officer on special duty (records) Nagaland Secretariat.

25 File no. pol/1247 of 1924, Aptt. Department, Political Branch, Record Cell, Nagaland Secretariat.

26 File no. pol/1299 of 1926, Aptt. Department, Political Branch, Record Cell, Nagaland Secretariat.

27 Annual Administrative report of the Mokokchung subdivision for the year 1920–21. File no 324. Memo no. 68, General Department, dated 13 May 1921. Nagaland State Archives.

28 Ibid., p. 14.

29 File No. Pol/1924 of 1924. Appointment and Political Department, Political Branch (Record Cell), Nagaland Secretariat.

30 File No. 08/305 of 1938, General Administrative Report of the Naga Hills District, 1937–39, Record Cell, Nagaland Secretariat.

31 Copy of letter No. 6053-G, dated 24 December 1939, from the Deputy Commissioner, Naga Hills, File no. 11K-1 7896 Agriculture Department, Assam. Assam Secretariat, Administrative (Records and Library), Dispur.

32 See, Annual Report of the Department of Agriculture, 1940, Assam Secretariat, Administrative (Records and Library), Dispur, p. 258.

33 Census of Assam, 1951, pp. 62–63.

34 See, the pamphlet 'How to Save the Soil in the Hills of Assam'. Written by R. C. Woodford and P. C. Khongwir, who were posted as agricultural inspectors in the Assam frontier. It appeared in *Assam Agricultural Bulletin* no. 2, 1939 and also appeared as a reference in C. R. Stonor and P. C. Khongwir's compilation in Annual Report of the Excluded and Partially Excluded Areas, Assam for the year ending 31 March 1947, p. 292.

35 This comes after my discussion with key informants during fieldwork (2008–9).

36 See the colonial debates and correspondence on hill administration and settlement of boundary question between Assam and Burma. IOR 19106/M/3/1461 and 1457.

37 Annual Report of the Department of Agriculture, 1946, Assam Secretariat, Administrative (Records and Library), Dispur, p. 247.

38 Ibid.

39 Ibid.

40 Annual Report of the Department of Agriculture, 1948, Assam Secretariat, Administrator (Records and Library) Department, Dispur, p. 245.

41 Ibid.

42 File No AF 482/54, 1954, Agriculture Department Forest Branch, Five Year Plan, Chapter XXI Forest Implementation and Recommendation. Record Cell, Department of Art and Culture, Nagaland.

43 NEPED stands for Nagaland Empowerment of People through Economic Development, a government-organized NGO that started in 1995 primarily to create new livelihood avenues for *jhum* farmers in Nagaland state.

44 This is based on my personal communication with people who collaborated with Malcolm Cairns in the project. I could not get in direct touch with Cairns as he has been permanently ill since 2009. I draw my observation reading other cross-references to his work in the NEPED resource handbook.

45 See also Puspanjali Deori's 2005 account, in which she details a number of incidents from colonial records showing that short-term successes and failures were a common outcome of colonial policy on land improvement followed by dedicated colonial civil servants, forest officials and agricultural department staff motivated by the idea of modernizing farming in Naga Hills.

5 Land and Land-Based Relations in a Yimchunger Naga Village: From a Book View to a Field View

1 The term 'low commoditized crop' has been used by Henry Bernstein in his study of African agriculture, where farmers grow certain commercial crops both for home consumption and for sale.

2 Leangkonger was referred to as the 'collection village' because it had households that had migrated from 17 different villages. The reasons for forming the new village were not clear from the villagers' descriptions. Some explained it as a result of clan conflict, while others attributed it to the outbreak of pandemics in the original villages.

3 As the village was formed by carving out land taken from three other villages (Siphonger, Shamator and Sanghpur) the clan members who set out to open up this new village had to pay gifts to the village heads.

4 The Yimchunger priest, and also the headman, who had the first right to sanctify the field before each cropping season – the first settler.

5 A classic example of colonial documentation of land relations can be found in Sir Keith Cantlie's *Notes on Khasi Law*. In this classic book, Cantlie gives a descriptive standardized account of colonial land assessment in the Khasi Hills, where colonial officials came to classify Khasi land tenureship based on their customary land use. The colonial classification was based on its matrix of the economic utility of land rather than the social relations that were embedded in land relations. A similar description of land use and relations can be found in the colonial classification of Naga land use in the monographs and official literature that was produced on Naga land relations.

6 These courts are attached to the deputy commissioner's office and function on customary dispute-settlement practice.

6 The Politics of Time: The Missionary Calendar, the Protestant Ethic and Labour Relations among the Eastern Nagas

1 In this chapter, the term 'church' refers not to the fellowship of the body of believers as it is theologically understood but to the institutionalized and bureaucratic structure called the church, comprising the personnel who wield the power to determine the making and nature of official church policies.

2 My interaction with a farmer in Leangkonger village, 11 April 2008.

3 The new tubers were introduced according to the villagers by the Ao missionaries. During fieldwork I enjoyed these tubers at pastime and during evening snack break and on Sundays after church prayers.

4 *Huker Baptist Church Golden Jubilee Khimtan Tsun*, January 1997, p. 16.

5 The village council contained a chairman, the *gaon burhas*, and the council members. All of these customary heads were below the Citizen, the overarching body. The Citizen was mainly responsible for addressing theft, petty crimes, abuses, imposing fines and defining membership.

6 Interview with a village farmer, 14 April 2009, Leangkonger village.

7 According to my household survey, that was carried out among 104 households, villagers' consumption expenditure data revealed that they had to spend a substantial portion of their disposable income in buying milk, tea and sugar.

7 Micro-Politics of Development Intervention: Village Patrons, Community Participation and the NEPED Project

1 Participatory rural appraisal is an approach used by non-governmental organizations and other agencies involved in international development. The approach aims to incorporate the knowledge and opinions of rural people in the planning and management of development projects and programmes.

2 These institutions are UN sister organizations, conservation NGOs, and specialized agricultural support institutions. The IIRP, UNEP–SENA and WWF became part of the project in its various phases.

3 See NEPED. 2007. 'Adding Value to Shifting Cultivation in Nagaland', vol. 2. *Nagaland Environment Protection through Economic Development.*

4 The difference in household numbers comes from the overrepresentation of population size in census data.

5 I have used 'Antham' as a pseudonym for my respondent to maintain anonymity and confidentiality.

6 There is a long-standing discussion on 'cultural brokers' in anthropological literature following Geertz (1960) and Wolf (1956). I follow Mosse and Lewis (2005) to define brokers as intermediaries in development projects–people who, by virtue of their education and social status, establish patronage with the village residents.

7 As revealed from my baseline household survey. The beneficiary community was thus very small and nucleated from the village community as conceived in development project.

8 I use household numbers to conceal the identities of my respondents as they could be under potential risk of being exposed. The village settlement map can be used to locate the relative position of the villagers in the study area. The readers can also relate to the morphology of the village and the nature of people who could become beneficiaries in the project, based on the local conditions.

9 I have used 'Nikon' as a pseudonym for my respondent to maintain anonymity and confidentiality.

10 Mini-hydroelectric power plant was successfully commissioned at Chizami village in Phek district in 2006 under the leadership of then MNES Secretary Gokhala. Funded by the Ministry of Non-Conventional Energy Sources, the 100×2 KW power plant was jointly set up by the village councils of Chizami Village and nearby Sumi Village with the technical assistance of Nagaland Renewable Energy Development Agency. Similar projects were planned and designed by the NEPED staff for other Naga villages.

8 Conclusion

1 Excerpts from interview with Asoba in the study village Leangkonger, 26 July 2009.

BIBLIOGRAPHY

Agamben, G. 2005. *State of Exception*. Chicago and London: University of Chicago Press.

Agrawal, A. and K. Sivaramakrishnan. 2000. 'Introduction'. In *Agrarian Environments: Resource, Representation and Rule in India*, edited by A. Agrawal and K. Sivaramakrishnan, 1–22. Durham, NC: Duke University Press.

Aier, A. and S. Changkija. 2003. 'Indigenous Knowledge and Management of Natural Resources'. In *Anthropology of North-East India*, edited by T. B. Subba and B. B. Ghosh, 333–78. Delhi: Orient Longman.

Aier, A. and J. Tiatoshi. 2009. 'Re-interpreting the Myth of Longterok'. *Indian Folklife*, 33: 5–9.

Ainslie, J. R. 1952. 'Resolution 10. "Shifting Cultivation", Report on the Committee on Shifting Cultivation'. Proceeding of the Empires Forest Conference, II, 386–87.

Allen, B. C. 1905. *Assam District Gazetteer, Naga Hills and Manipur*. Vol. 9. Calcutta: Baptist Mission Press.

Angami, Z. 2008. *Nagaland Village Empowerment Laws*. Kohima: Novelty Printing Press.

Anonymous. 2007. *Village Guard Nagaland Souvenir*, 50th Anniversary Celebration (1957–2007). Kohima: Government of Nagaland.

Ao, A. 1970. *A Brief Historical Account of Nagaland*. Kohima: Naga Institute of Culture.

Appadurai, A. 2003 "The Capacity to Aspire," in *Culture and Public Action*, edited by V. Rao and M. Walton, California: Stanford University Press.

Arya, A. 2004. *The Land of the Nagas*. Mipan. Ahmedabad.

Balfour, H. and J. H. Hutton. 1938. 'Through the Unexplored Mountains of Assam-Burma Border'. *The Geographical Journal* 91, no. 3: 216–19.

Barnes, A., P. C. W. Harumi Befu, A. Gutkind, Weingrod, N. E. Whitten, Jr. and P. Hirch. 1989. 'Forest, Forest Reserve, and Forest Land in Thailand'. *The Geographical Journal* 156, no 2: 166–74.

Barpujari, H. K.1981. *Problems of the Hill Tribes: North East Frontier (1873–1962)*. Vol. 3. Guwahati: Spectrum Books.

———. 1998. *North-East India: Problems, Policies and Prospects*. Delhi: Spectrum Publications.

Barpujari, S. K. 2003. *The Nagas: The Evolution of Their History and Administration, 1832–1939*. Delhi: Spectrum Publications.

Baruah, S. 2001. 'Clash of Resource Use Regimes in Colonial Assam: A Nineteenth-Century Puzzle Revisited'. *Journal of Peasant Studies* 28: 3109–24.

———. 2003a. 'Confronting Constructionism: Ending India's Naga War'. *Journal of Peace Research* 40, no. 3: 321–38.

———. 2003b. 'Nationalising Space: Cosmetic Federalism and the Politics of Development in Northeast India'. *Development and Change* 34, no. 5: 915–39.

———. 2005. *Durable Disorder: Understanding the Politics of North East India.* Delhi: Oxford University Press.

———. 2007. 'Postfrontier Blues: Toward a New Policy Framework for Northeast India'. *Policy Studies* 33: 1–83.

Baviskar, A. 2004. 'Between Micro politics and Administrative Imperatives: Decentralization and the Watershed Mission in Madhya Pradesh, India', *European Journal of Development Research*, 16: 26–40.

Baviskar, B. S. 2002. 'Walking on the Edge of Factionalism: An Industrial Co-operative in Rural Maharashtra'. In *The Fieldworker in the Field*, edited by M. N. Srinivas, A. M. Shah and E. A. Ramaswamy, 184–200. Delhi: Oxford University Press.

Bernstein, H. and P. Woodhouse. 2001. 'Telling Environmental Change Like It Is? Reflections on a Study in Sub-Saharan Africa'. *Journal of Agrarian Change* 1, no. 2: 283–324.

Berreman, G. 1972. 'Prologue: Behind Many Masks: Ethnography and Impression Management'. In *Hindus of the Himalayas: Ethnography and Change*, edited by G. D Berreman, xvii–lvii. Berkeley: University of California Press.

Berry, S. 1988a. 'Concentrating without Privatisation? Some Consequences of Changing Patterns of Rural Land Control in Africa'. In *Land and Society in Contemporary Africa*, edited by R. E. Downs and S. P. Reyna, 53–75. Hanover, NH, and London: University Press of New England.

———. 1988b. 'Property Rights and Rural Resource Management: The Case of Tree Crop in West Africa'. *In Cahiers d' Orstom Science Humains* 24, no. 1: 3–16.

———. 1989. 'Access, Control and Use of Resources in African Agriculture: An Introduction', *Africa: Journal of the International African Institute* 59, no. 1: 1–5.

———. 1992. 'Hegemony on a Shoestring: Indirect Rule and Access to Agricultural Land'. *Africa: Journal of the International African Institute* 62, no. 3: 327–55.

———. 1993. *No Condition Is Permanent, The Social Dynamics of Agrarian Change in Sub-Saharan Africa.* Madison: University of Wisconsin Press.

———. 2004. 'Reinventing the Local? Privatisation, Decentralisation and the Politics of Resource Management'. *African Study Monographs* 25, no. 2: 79–101.

Blackburn, S. 2003. 'Memories of Migration: Notes on Legends and Beads in Arunachal Pradesh'. *European Bulletin of Himalayan Research* 25 and 26: 1–60.

Bloch, M. and J. Parry, eds. 1982. *Death and the Regeneration of Life.* Cambridge: Cambridge University Press.

Boomgaard, P. and D. Hanley. 2004a. *Smallholders and Stockbreeders: Histories of Food Crop and Livestock Farming in Southeast Asia.* Leiden: KVLT Press.

———. 2004b. 'Agricultural and Livestock Histories of Southeast Asia: Approaches and Perspectives'. In *Smallholders and Stockbreeders: Histories of Foodcrop and Livestock Farming in Southeast Asia*, edited by P. Boomgaard and D. Hanley, 1–18. Leiden: KVLT Press.

Booth, A. 1988. *Agricultural Development in Indonesia.* Sydney: Allen and Unwin.

Booth, N. S. Jr. 1975 'Time and Change in African Traditional Thought'. *Journal of Religion in Africa* 7, no. 2: 81–91.

Borneman, J. 2009. 'Fieldwork and Punishment in Lebanon'. In *Violence: Ethnographic Encounters*, edited by P. Ghassem-Fachandi, 108–30. Oxford and New York: Berg.

Boserup, E. 1965. *The Condition of Agriculture Growth: The Economics of Agrarian Change Under Population Pressure.* London: Allen and Unwin.

Botkin, D. B. 1990. *Discorded Harmonies: A New Ecology for the Twenty First Century.* Oxford: Oxford University Press.

Bourdieu, P. 1968. 'The Attitude of the Algerian Peasant Towards Time'. In *Mediterranean Countryman: Essays on the Social Anthropology of Mediterranean*, edited by J. Pitt-Rivers. Paris: Mouton.

———. 1977. *Outline of a Theory of Practice*. Cambridge: Cambridge University Press.

Bourgois, P. 1991 [2007]. 'Confronting the Ethics of Ethnography: Lessons from Fieldwork in Central America'. In *Decolonising Anthropology: Moving Further Towards an Anthropology for Liberation*, edited by F. V. Harrison, 110–26. Washington, DC: American Anthropological Association.

Brow, J. and J. Weeramunda. 1992. *Agrarian Changes in Sri Lanka*. New Delhi: Sage.

Bruce, J. W. 1998. 'Review of Tenure Terminology'. *Tenure Brief (Land Tenure Center)*, no. 1, July.

Bruce, J. W. and R. Noronha. 1985. 'Land Tenure Issues in the Forestry and Agro-Forestry Project Context'. Proceedings of an International Workshop on Tenure Issues in Agroforestry, 27–31 May, Nairobi.

Bryant, R. L. 1997. *The Political Ecology of Forestry in Burma 1824–1944*. London: Hurst.

Burling, R. 1965. *Hill Farms and Paddy Fields: Life in Mainland South East Asia*. Englewood Cliffs, NJ: Englewood Cliffs.

Butler, J. 1875. 'Rough Notes on the Angami Nagas', *Journal of the Asiatic Society of Bengal* 44: 4, 329–32.

Butler. H. 1926. "Slavery in Burma: Hakwang Valley Cleared", *The Times*, 26 June.

Cairns, M. 2007. 'The Alder Managers: The Cultural Ecology of a Village in Nagaland, N. E. India'. Ph.D. Thesis in Anthropology, The Australian National University, Canberra.

Cairns, M. and H. Brookfield. 2011. 'Composite Farming System in an Era of Change: Nagaland: North-east India'. *Asia Pacific Viewpoint* 52, no. 1: 55–84.

Cairns, M., S. Keitzar and T. A. Yaden. 2007a. 'Shifting Forest in Northeast India: Management of *Alnus nepalensis* as an Improved Fallow in Nagaland'. In *Voices from the Forest: Integrating Indigenous Knowledge into Sustainable Upland Farming*, edited by M. Cairns, 341–78. Washington, DC: Resources for the Future.

Campbell, J. 2011. 'The Problem of Ethics in Contemporary Anthropological Research', *Anthropology Matters* 10, no. 1: 1–17.

Cantlie, K. 2008 [1974]. *Notes on Khasi Law*, edited by A. S. Khongphai. Shillong: Khongphai.

Cederlof, G. 2010. *Landscape and the Law: Environmental Politics, Regional Histories, and Contest over Nature*. New Delhi: Permanent Black.

Chanock, M. 1991. 'Paradigms, Policies and Property. A Review of Customary Law of the Land Tenure'. In *Law in Colonial Africa*, edited by K. Mann and R. Roberts, 61–84 Portsmouth: Heineman.

Chatterjee, I. 2002. 'Territory and Identity: Slaves during the Age of Imperial Frontier Management'. In *Space, Territory and State: New Readings in International Politics*, edited by R. Samaddar, 63–86. Delhi: Orient Longman.

———. 2006. 'Slavery, Semantics and the Sound of Silence'. In *Slavery and South Asian History*, edited by I. Chatterjee and R. M. Eaton, 287–316. Bloomington: Indiana University Press.

Chayanov, A. V. 1966 [1926]. *The Theory of Peasant Economy*. Homewood, IL: Richard D. Irwin.

Chen, K-H. and D. Morley. 1996. 'On Postmodernism and Articulation'. In *Stuart Hall: Critical Dialogues in Cultural Studies*, edited by K-H. Chen and D. Morley, 131–50. London and New York: Routledge.

Chena, S. M. 1992. 'Introduction' in *Nagaland: A Contemporary Ethnography*, edited by S. M. Chena. New Delhi: Cosmo.

Cohn, B. S. 1996. *Colonialism and Its Forms of Knowledge: The British in India*. Delhi: Oxford University Press.

Condominas, G. 2009. 'Anthropological Refection on Swidden Change in South-east Asia'. *Human Ecology* 37: 265–67.

Conklin, H. C. 1961 [1957]. *Hanunoo Agriculture: A Report on Integral System of Shifting Cultivation in the Philippines*. Rome: Food and Agricultural Organisation of the United Nations.

Cook, B. and U. Kothari. 2001. *Participation: The New Tyranny?* London: Zed Books.

Daniel. V. E. 1996. *Charred Lullabies: Chapters in an Anthropography of Violence*. Princeton: Princeton University Press.

Das, D. 2003. 'Trauma and Testimony: Implications for Political Community'. *Anthropological Theory* 3, no. 3: 293–307.

———. 2009. 'Colonial Construction of a Frontier Space: Debating the Inner Line Regulation in the Sibsagar-Naga Hills Frontier'. Paper read in the Writing the North East: New Historical Perspectives seminar held at Jawaharlal Nehru University (Centre for Historical Studies), 14–16 January 2009, New Delhi.

Das Gupta, U. and D. Chattopadhyay, eds. 2010. *Science and Modern India: An Institutional History, c 1784–1947, Philosophy and Culture in Indian Civilisation*, Delhi: Pearson.

Deori, P. 2005. *Environmental History of the Naga Hills 1881–1947*. Delhi: Akansha Publishing House.

DONER. 1996. 'North Eastern Region Vision 2020'. Shillong: Ministry of Development of North Eastern Region and North Eastern Council, 1–293.

Douglas, M. and B. Isherwood. 1978. *The World of Goods: Towards an Anthropology of Consumption*. London: Rutledge.

Dove, M. R. 1983. 'Theories of Swidden Agriculture, and the Political Economy of Ignorance'. *Agroforestry Systems* 1, no. 2: 85–99.

———. 1985. 'The Agroecological Mythology of the Javanese and the Political Economy of Indonesia'. *Indonesia* 39: 1–36.

———. 1993. 'A Revisionist View of Tropical Deforestation and Development'. *Environmental Conservation* 20, no. 1: 17–56.

Eaton, R. M. 1997. Comparative History as World History: Religious Conversion in Modern India. *Journal of World History* 8, no. 2: 243–71.

Elaide, M. 1957. *Cosmos and History*. New York: Harper.

Ellen, R. 1999. 'Forest Knowledge, Forest Transformation: Political Contingency, Historical Ecology and Recognition of Nature in Central Seram', edited by T. M. Li, 131–57. Amsterdam: Harwood.

Elwin, V. 1956. *Policy Dossier 'New Deal' for the North East*. Shillong: NEFA Administration.

———. 1961. *Nagaland*. Shillong: NEFA Administration.

———. 2007 [1939]. *The Baigas*. Bombay: Oxford University Press.

Fabian, J. 1983. *Time and the Other: How Anthropologist Makes Its Objects*. New York: Columbia University Press.

Fairhead, J. and M. Leach. 1996. *Misreading the African Landscape: Society and Ecology in a Forest-Savannah Mosaic*. Cambridge: Cambridge University Press.

Faminov, Marley D. and K. K. Klein. 2001. 'On Farm Testing and Dissemination of Agro-forestry Among Slash-and-Burn Farmers in Nagaland'. *Development and Practice*, 11, 471–86.

FAO. 1957. 'Shifting Cultivation.' *Unasalva* 11: 9–11.

———. 2006. *The State of the World's Forests*. Rome: FAO.

———. 2009. *State of the World's Forests*. Rome: FAO Publications.

Feder, G. and D. Feeny.1989. 'Land Tenure and Property Rights: Theory and the Decline of Property Rights in Man in Thailand, 1800–1913'. *Journal of Economic History* 49: 285–96.

Feldman, A. 1991 [1994, 1997]. *Formations of Violence: The Narrative of the Body and Political Terror in Northern Ireland*. Chicago: University of Chicago Press.

———. 1995. 'Ethnographic States of Emergency'. In *Fieldwork Under Fire: Contemporary Studies in Violence and Survival*, edited by C. Nordstrom and A. Robben, 224–53. Berkeley: University of California Press.

Fenny, D. 1989. 'The Decline of Property Rights in Man in Thailand, 1800–1913'. *The Journal of Economic History* 49, no. 2: 285–96.

Fiskesjö, M. 2011. 'Slavery and the Commodification of People: "Wa" Slaves and their Chinese "Sisters"', *Focaal: Journal of Global and Historical Anthropology* 59: 3–18.

Forsyth, T. and A. Walker. 2008. *Forest Guardians: Forest Destroyers: The Politics of Environmental Knowledge in Northern Thailand*. Seattle and London: University of Washington Press.

Foucault, M. 1977. *Discipline and Punish: The Birth of the Prison*. New York: Random House.

Franke, M. 2009. *War and Nationalism in South Asia: The Indian State and the Nagas*. New York: Routledge.

Friedman, H. 1980. 'Household Production and the National Economy: Concepts for the Analysis of Agrarian Formation'. *Journal of Peasant Studies* 7, no. 2: 158–84.

Friedman, J. 1985. 'Post-structuralism and the New Moon'. *Ethnos* 1–2, 123–33.

Fürer-Haimendorf, C. V. 1937. "Head-hunters Never Before Visited by Europeans. Among the Wild Naga Tribes in a Hitherto Unexplored Region of the Assam–Burma Border: Adventures with a British Expedition Sent to Suppress Slave-Raiding and Human Sacrifice". *Illustrated London News* 26 June.

———. 1938a. 'Through the Unexplored Mountains of the Assam–Burma Border'. *Geographical Journal* 91 no. 3: 201–16.

———. 1938b. "The Last Head-Hunting Feast of the Konyak Nagas of Assam". *Illustrated London News* 5 February.

———. 1939a. 'Konyak Naga Head-hunters as Agriculturist', and 'Irrigation Engineering Marvels of the Primitive Angami Nagas'. *Illustrated London News*, 8 July.

———. 1939b. 'The Life of Women Among a People Untouched by Civilisation: Domestic Activities in Villages of Headhunting Nagas'. *Illustrated London News*, 9 July.

———. 1947. *Tour Diaries of a Special Officer in Subansari 1944–1945*. Calcutta: Baptist Mission Press.

———. 1955. *Himalayan Barbary*. London: John Murray.

———. 1969 [1939]. *The Konyak Nagas: An Indian Frontier Tribe*. London and New York: Holt, Rinehart and Winston.

Gadgil, M. and G. Ramachandra. 1993. *This Fissured Land: An Ecological History of India*. Oxford: Oxford University Press.

Gangte, T. S. 2006. 'Struggle for Identity and Land among the Hill Peoples of Manipur'. Paper Presented at National Conference on Folklore, Identities and Performance. Imphal, 23 May.

Ganguli, M. 1984. *A Pilgrimage to the Nagas*. Delhi: Oxford IBH.

Geertz, C. 1963. *Agricultural Involution: The Process of Ecological Change in Indonesia*. Berkeley: University of California Press.

———. 1973. *Interpretation of Cultures: Selected Essays*. New York: Basic Books.

George, K. M. 1996. *Showing Signs of Violence: The Cultural Politics of a Twentieth-Century Headhunting Ritual.* Berkeley: University of California Press.

Ghassem-Fachandi, P. 2009. 'Introduction'. In *Violence: Ethnographic Encounter*, edited by P. Ghassem-Fachandi. Oxford and New York: Berg.

Gilmartin, D, 1994. 'Scientific Empire and Imperial Science: Colonialism and Irrigation Technology in the Indus Basin'. *The Journal of Asian Studies* 53, no. 4 (November): 1127–49.

Gluckman, M. 1963. *Customs and Conflicts in Africa.* Oxford: Blackwell.

Godwin-Austin, H. H. 1873. *Topographical Survey of India.* Calcutta: Baptist Missionary Press.

———. 1875. 'On the Rude Stone Monuments of Certain Naga Tribes with Some Remarks on the Customs'. *Journal of the Royal Anthropological Institute* 4: 144–47.

Goffman, E. 1959. *The Presentation of Self in Everyday Life.* New York: Penguin.

Goodenough, W., J. H. Hutton, W. Rich and F. Kingdon-Ward. 1937. 'From the Brahmaputra to the Chindwin: Discussion'. *The Geographical Journal* 89, no. 4: 323–26.

Gray, A. 1986. 'The British in Nagaland: Their Anthropology and Their Legacy'. In *The Naga Nation and Its Struggle Against Genocide*, Document 56, Copenhagen: IWGIA.

Green, M. 2000. 'Participatory Development and the Appropriation of Agency in Southern Tanzania'. *Critique of Anthropology* 20, no. 1: 67–89.

Grigg, D. 1982. *The Dynamics of Agrarian Change: The Historical Experience.* London: Hutchinson.

Grove, R. H. 1998. *Ecology, Climate and Empire: The Indian Legacy of Global Environmental History 1400–1940.* Oxford: Oxford University Press.

———. 1995. *Green Imperialism: Colonial Expansion, Tropical Island Edens and the Origins of Environmentalism 1600–1860.* Cambridge: Cambridge University Press.

Guha, S. 1999. *Environment and Ethnicity in India 1200–1900,* Cambridge: Cambridge University Press.

Gunasinghe, N. 1992. 'Transforming the Trajectories of Agrarian System in Jaffna and Newara Eliya Districts'. In *Agrarian Change in Sri Lanka,* edited by J. Brow and J. Weeramunda, 201–25. New Delhi: Sage.

Gurani, S. 1995. 'Regimes of Control, Strategies of Access: Politics of Forest Use in Uttarakhand Himalayas, India'. In *Agrarian Environment: Resource, Representation and Rule*, edited by A. Agrawal and K. Sivaramakrishnan, 170–90. Durham, NC: Duke University Press.

Hames, R. 2007. 'Ecologically Noble Savage Debate'. *Annual Review of Anthropology* 36: 177–90.

Handique, R. 2004. *British Forest Policy in Assam.* New Delhi: Concept Publishing Company.

Hafner, Robert, W. 1993. 'World Building and the Rationality of Conversion'. *In Conversion to Christianity: Historical and Anthropological Perspectives on a Great Transformations*, edited by Robert W. Hafner, 3–44. Berkley: University of California Press.

Hann, C. M. 1998. 'Introduction: The Embeddedness of Property'. In *Property Relations: Renewing the Anthropological Tradition*, edited by C. M. Hann. Cambridge: Cambridge University Press.

Hardiman, D. 1981. *The Coming of the Devi: Adivasi Assertion in Western India.* Oxford: Oxford University Press.

———. 2008. *Missionaries Medicine: A New Modernity for Tribal India.* Manchester: Manchester University Press.

Harrison. M. 1977. 'The Peasant Mode of Production in the Work of A. V. Chayanov'. *Journal of Peasant Studies* 4, no. 4: 323–36.

Hart, G., E. Turton and B. White, eds. 1989. *Agrarian Transformations: Local Processes and the State in South Asia.* Berkeley: University of California Press.

Hayami, Y. and T. Kawahgeo. 1981. *Asian Village Economy at the Crossroads: An Economic Aspect to Institutional Change*. Tokyo: University of Tokyo Press.

Haymes, R. 2007. 'The Ecologically Nobel Savage Debate'. *The Annual Review of Anthropology* 36, no. 7: 177–90.

Henry, F. 1966. 'The Role of the Fieldworker in an Explosive Political Situation'. *Current Anthropology* 7, no. 5: 552–59.

Hobart, Mark. 1993. 'Introduction: The Growth of Ignorance'. In *An Anthropological Critique of Development: The Growth of Ignorance?*, edited by Mark Hobart. London and New York: Routledge, pp 1–30

Hobsbawm, E. and T. Ranger. 1983. *The Invention of Tradition*. Cambridge: Cambridge University Press.

Hodge, J. M. 2007. *Triumph of the Expert: Agrarian Doctrines of Development and the Legacies of British Colonialism*. Athens, OH: Ohio University Press.

Hodson, T. C. 1911. *The Naga Tribes of Manipur*. London: Macmillan.

Hoebel, E. A. 1966. *Anthropology: The Study of Man*. New York: McGraw-Hill.

Hoskins, J. 1993. *The Play of Time: Kodi Perspective of Calendar, History and Exchange*. Berkeley and Los Angles: University of California Press.

———. 1996. *Headhunting and the Social Imagination of South Asia*. Stanford: Stanford University Press.

Howell, N. 1990. *Surviving Fieldwork: A Report of the Advisory Panel on Health and Safety in Fieldwork*. Washington, DC: American Anthropological Association.

Hutton, J. H. 1920. 'Leopard Men in the Naga Hills'. *The Journal of the Royal Anthropological Institute of Great Britain and Ireland* 50: 41–51.

———. 1921a. *The Angami Nagas*. London: Macmillan.

———. 1921b. *The Sema Nagas*. London: Macmillan.

———. 1928. 'The Significance of Head-Hunting in Assam'. *The Journal of the Royal Anthropological Institute of Great Britain and Ireland* 58: 399–408.

———. 1945. 'Problems of Reconstruction in the Assam Hills'. *Journal of Royal Anthropological Institute of Great Britain and Ireland* 75, nos. 1 and 2: 1–7.

———. 2008 [1929]. *Diaries of Two Tours in the Unadministered Area East of Naga Hills: Memoirs of the Asiatic Society of Bengal*. Vol. 11, no 1. Delhi: Mittal.

Iijama. I. 1970. 'Socio Cultural Change among the Shifting Cultivators through the Introduction of Wet Rice Cultivation A Case Study of Kerens in Northern Thailand', Memoirs of the College of Agriculture. Japan: Kyoto University, 79, no. 3: 1–35.

Iralu, K. D. 2003. *Nagaland and India: The Blood and the Tears. A Historical Account of the Fifty-Two Year Indo–Naga War and the Story of Those Who Were Never Allowed to Tell It*. Kohima: Author's private publication.

———. 2002. 'The Fifty-Four Year Indo–Naga Conflict: A Question of Internal Indian Ethnic Conflict or a Conflict between Two Nations'. A Paper Presented at the Indian Council of Social Sciences Research (ICSSR) Sponsored by the National Seminar on Resolving Ethnic conflicts in North East India. Guwahati: 11–12 November.

Ives, J. 1987. 'The Theory of Himalayan Environmental Degradation: Its Validity and Application Challenged by Recent Research'. *Mountain Research and Development* 7, no. 3: 189–99.

Ives, J. D. and P. Ives. 1989. *The Himalayan Dilemma: Reconciling Development and Conservation*. London and New York: Routledge.

Jacobs, J., A. Macfarlane, S. Harrison and A. Herle. 1990. *The Nagas: Hill Peoples of Northeast India: Society, Culture, and the Colonial Encounter*. London: Thames and Hudson.

Jamir, T. and D. Vasa. 2008. 'An Archaeological Investigation at Chungliyimti, Tuensang District'. Unpublished Project Report: Anthropological Society of Nagaland.

Kar, B. 2009. 'When Was the Postcolonial? A History of Policing Impossible Lines'. In *Beyond Counter-insurgency: Breaking the Impasse in Northeast India*, edited by S. Baruah. Delhi: Oxford University Press.

Karlsson, B. G. 1997. *Contested Belonging: An Indigenous People's Struggle for Forest and Identity in Sub-Himalayan Bengal.* Lund: Lund University.

———. 2011. *Nature and the Nation: Politics of Nature in North East India.* Delhi: Orient Blackswan and Social Science Press.

Kauffmann, H. E. 1939. 'Kurze Ethnographie der nordlichen Sangthan Naga (Lophomi) Assam', *Anthropos* 34.

Kejariwal, O. P. 1987. 'The Indian History Congress and Historical Research in North East India'. In *Proceedings of the North-East History Association*, 33–81. Kohima: North-East India History Association.

Kovats-Bernat, J. C. 2002. 'Negotiating Dangerous Fields: Pragmatic Strategies for Fieldwork Amid Violence and Terror'. *American Anthropologist* 104, no. 1: 208–22.

Kumar, B. B. 1996. *Re-organisation of North-East India (Facts and Documents).* New Delhi: Omsons Publications.

———. 2009. 'Alienation and Insurgencies in the North-East India', *Dialogue.* http://www.asthabharati.org/Dia_Oct%2009/B.B.html. Accessed on 12.01.2011.

Kumar, N. 1992. *Friends, Brothers, Informants: Fieldwork Memories of Beneras.* Berkeley: University of California Press.

Langkua, M. 2004. 'The *Jhum* System in Nagaland under Population Pressure'. Unpublished MA Dissertation submitted in the Department of Institute for Silviculture, Georg-August University, Göttingen.

Leach, E. R. 1954. *The Political System of Highland Burma: A Study of Kachin Social Structure.* London: Bell. (Originally LSE Monograph on Social Anthropology no. 44).

———. 1961. *Rethinking Anthropology.* London: Anthlon Press.

———. 1960. 'The Frontiers of "Burma"'. *Comparative Studies in Society and History*, 3, no. 1: 49–68.

Leach, M. and R. Mearns, 1996. *The Lie of the Land: Challenging Received Wisdom on the African Environment.* Portsmouth, NH: Heinemann.

Leigh Pigg, S. 1993. 'Inventing Social Categories through Place: Social Representation of Development in Nepal'. *Contemporary Studies in Society and History* 34, no. 3: 491–513.

Li, T. M. 1996a. 'Household Formation, Private Property, and the State'. *Sajourn* 11, no. 2: 259–87.

———. 1996b. 'Images of Community: Discourse and Strategy of Property Relations'. *Development and Change* 27, no. 3: 501–27.

———. ed. 1996c. *Transforming the Indonesian Upland: Marginality, Power and Production.* Amsterdam: Harwood.

———. 1997. 'Producing Agricultural Transformation at an Indonesian Periphery'. In *Economic Analysis beyond the Local System: Monograph on Economic Anthropology*, edited by R. E. Blanton, P. N. Peregrine, D. Winslow and T. D. Hall. Maryland: University Press of America.

———. 1998. 'Working Separately but Eating Together: Personhood, Property and Power in Conjugal Relations'. *American Ethnologist* 25, no. 4: 675–94.

———. 2000. 'Articulating Indigenous Identity in Indonesia: Resource Politics and the Tribal Slot'. *Comparative Studies in Society and History* 42, no. 1: 149–79.

———. 2007a. 'Adat in Central Salwasi: Contemporary Deployment'. In *The Reinvention of Tradition in Indonesian Politics: The Deployment of Adat from Colonialism to Indigenism*, edited by J. S. Davidson and D. Hanley. London and New York. Routledge.

———. 2007b. *The Will to Improve: Governmentality, Development, and the Practice of Politics.* Durham, NC: Duke University Press.

———. 2010. 'Agrarian Class Formation in Upland Sulawesi 1990–2010', the Challenge of the Agrarian Transition in Southeast Asia Working Paper No. 9. Montreal: Canadian Research Chair in Asian Studies, 1–44.

Lobo, L. 1990, 'Becoming a Marginal Native'. *Anthropos* 85: 125–38.

Lonkumer, A. 2010. *Reform, Identity and Narratives of Belonging: The Heraka Movement in North East India.* London and New York: Continuum Publication House.

Lotha, A. 2007. *A History of Naga Anthropology (1832–1947).* Dimapur: Chumpo Museum Publications.

———. 2009. 'Rearticulating Naga Nationalism', thesis presented in partial fulfilment of a PhD in Anthropology submitted at New York University, Anthropology Department.

Luithui, L. and N. Haskar. 1984. *Nagaland File: A Question of Human Rights.* Delhi: Lancer International.

Lynch, O. and E. Harwell. 2001. *Whose Resources? Whose Common Good? Towards an Environmental Justice Paradigm in Indonesia's Natural Resource Management.* Jakarta: ELSAM.

Maaker, E. De and V. Joshi. 2007. 'Introduction: The Northeast and Beyond: Region and Culture'. *South Asia: Journal of South Asian Studies* 30, no. 3: 381–90.

Mackenzie, A. M. 1979 [1884]. *The North Eastern Frontier of India.* New Delhi: Mittal Publications.

Marcea, E. 1965. *The Myth of the Eternal Return: Cosmos and History.* New York: Pantheon.

Mankekar, D. R. 1967. *On the Slippery Slope in Nagaland.* Bombay: Mankeda.

McLean, A. and A. Leibing. 2007. *The Shadow Side of Fieldwork: Exploring the Blurred Borders between Ethnography and Life.* London and New York: Blackwell Publishers.

Means, G. P. 2000. 'Human Sacrifice and Slavery in the "Unadministered" Areas of Upper Burma During the Colonial Era'. *Sojourn* 15, no. 2: 181–221.

Meek, C. K. 1949. *Land Laws and Customs in the Colonies.* London: Oxford University Prsess.

Mertz, O. and S. J. Leisz, A. Heinemann, T. Kanok Rerkasem, and W. Dressler, V. C. Pham, K. C. Vu, D. Schmidt-Vogt, C. J. P. Colfer, M. Epprecht, C. Padoch and L. Potter. 2009. 'Who Counts? Demography of Swidden Cultivators in Southeast Asia'. *Human Ecology* 37: 281–89.

Midgal, J. 1974. *Peasant, Politics and Revolution.* Princeton: Princeton University Press.

Miers, S. 2003. *Slavery in the Twentieth Century: The Evolution of a Global Problem.* Oxford: Oxford University Press.

Mills, J. P. 1922. *The Lotha Nagas.* London: Macmillan.

———.1926a. 'Certain Aspects of Naga Culture'. *The Journal of the Royal Anthropological Institute of Great Britain and Ireland* 56: 27–35.

———. 1926b. *The Ao Nagas.* London: Macmillan.

———.1926 and 1927. *Tour Diary in the Chittagong Hill Tracts.* Calcutta: Baptist Mission Press.

———.1995. *The Pangsha Letters: An Expedition to Rescue Slaves in the Naga Hills.* Oxford: Pitt Rivers Museum.

Misra, S. 1998. 'The Nature of Colonial Intervention in the Naga Hills, 1840–80'. *Economic and Political Weekly* 33, no. 51: 3273–79.

Mitra, K. and R. Gupta. 2009. 'Indigenous Peoples' Forest Tenure in India'. In *Land and Cultural Survival: The Communal Land Rights of Indigenous Peoples in Asia*, edited by J. Perera, 193–12. Manila: Asian Development Bank.

Moerman, M. 1968. *Agricultural Changes and Peasant Choices in a Thai Village.* Berkeley: University of California Press.

Moore, Hernietta and Vaughan, Megan. 1994. *Cutting Down Trees: Gender, Nutrition and Agricultural Change in the Northern Province of Zambia, 1890–1990*. Cambridge: Cambridge University Press.

Morris, B. 1982. *Forest Traders: A Socioeconomic Study of the Hill Pandaram* (London School of Economics Monograph in Social Anthropology, no. 55). London and New York: The Athlone Press.

Mosse, D. 1994. 'Authority, Gender and Knowledge: Theoretical Reflections on the Practice of Participatory Rural Development'. *PLA Notes* 24: 27–33.

———. 2005. *Cultivating Development: An Ethnography of Aid Policy and Practice*. London: Pluto Press.

———. 2006. 'Anti-social Anthropology? Objectivity, Objection, and the Ethnography of Public Policy and Professional Communities'. *Journal of the Royal Anthropological Institute* 12, no. 4: 935–56.

Nash, J. 1976. 'Ethnology in Revolutionary Setting'. In *Ethics and Anthropology: Dilemmas in Field Work*, edited by M. Rynkiewich and J. Spradley, 144–66. New York: Wiley.

Negi, A. K., B. P. Bhatt, N. P. Todaria and A. Saklani. 1997. 'The Effect of Colonialism on Forest and the Local People in Garwal Himalayas, India'. *Mountain Research and Development* 17, no. 2: 159–68.

NEPED. 1999. 'Building upon Traditional Agriculture in Nagaland'. NEPED and IIRP: Nagaland and Philippines.

Newell, S. B., Jr. 1975. 'Time and Change in African Traditional Thought'. *Journal of Religion in Africa* 7. Fasc. 2.

Ngo, T. T. T. 2016. *The New Way: Protestantism and the Hmong in Vietnam*, Seattle: University of Washington Press,

Nilan, P. 2002. '"Dangerous fieldwork" Re-examined: The Question of Researcher Subject Position'. *Qualitative Research* 2, no. 3: 363–83.

Nongbri, T. 2008. *Gender, Matriliny and Entrepreneurship: The Khasis of Northeast India*. Dehi: Zubaan.

Nordstrom, C. and J. Martin. 1992. 'The Culture of Conflict: Field Reality and Theory'. In *The Paths to Domination, Resistance and Terror*, edited by C. Nordstrom and J. Martin. Berkley, Los Angles and Oxford: University of California Press.

O'Brien, W. E. 2002. 'The Nature of Shifting Cultivation: Stories of Harmony, Degradation and Redemption'. *Human Ecology* 30, no. 4: 483–502.

Ohnuki-Tierney, E. 1973. 'Sakhalin Ainu Time Reckoning'. *Man*, New Series, 8, no. 2: 285–99.

Oli, M., P. Christine, F. Jefferson, R. A Cramb, S. J. Leisz, T. L. Nguyen and D. V. Tran. 2009. 'Swidden Changes in Southeast Asia: Causes and Consequences', *Human Ecology* 37: 259–64.

Oppitz, M., W. Marion and K. Thomas 2008. *Naga Identities: Changing Local Cultures in the Northeast of India*. Leiden: Snoeck, Brill.

Padel, F. 2009. *Sacrificing People: Invasions of a Tribal Landscape*. Delhi: Orient Black Swan. 2nd ed. [1995] of *The Sacrifice of Human Being: British Rule and the Konds of Orissa*.

Padoch, C., K. Coffey, O. Mertz, S. Leisz, J. Fox and R. Wadley. 2007. 'The Demise of Swidden in Southeast Asia? Local Realities and Regional Ambiguities'. *Danish Journal of Geography* 107, no. 1: 29–42.

Panini, M. N. 1991. *From the Female Eye: Accounts of Women Fieldworkers Studying Their Own Community*. Delhi: Hindustan.

Peluso, Nancy Lee. 1992. Rich Forests, Poor People: Resource Control and Resistance in Java. Berkeley, CA: University of California Press.

Peluso, N. L., and P. Vandergeest. 2001. 'Genealogies of the Political Forest and Customary Rights in Indonesia, Malaysia and Thailand'. *Journal of Asian Studies* 60: 3761–812.

Pelzer, K. J. 1945. *Pioneer Settlement in the Asiatic Tropics: Studies in Land Utilisation and Agricultural Colonisation in Southeast Asia*. New York: Institute of Pacific Relations.

Philip, P. T. 1983. *The Growth of the Baptist Church in Nagaland*. Guwahati: Christian Literature Centre.

Polanyi, K. 2001. *The Great Transformation*. Boston: Beacon Press.

Prasad, A. 1994. 'Forest and Subsistence in Colonial India: A Study of Central Province'. Unpublished PhD thesis, CHS, Jawaharlal Nehru University, New Delhi.

Pruett, G. 1974. 'Christianity, History, and Culture in Nagaland'. *Contributions to Indian Sociology* 8: 51–65.

Puthenpurakal, J. 1984. *Baptist Missions in Nagaland*. Shillong: Vendrame Missiological Institute.

Rai, M. B. 1956. 'Tour Report of the Tuensang Frontier Division'. Record Cell, Arunachal State Secretariat Archives.

Rajan, R. 1998a. 'Foresters and the Politics of Colonial Agroecology: The Case of Shifting Cultivation and Soil Erosion, 1920–50'. *Studies in History* 14, no. 2: 217–36.

———. 1998b. 'Imperial Environmentalism or Environmental Imperialism? European Forestry, Colonial Foresters and the Agendas of Forest Management in British India, 1800–1900'. In *Nature and the Orient: The Environmental History of South and Southeast Asia*, edited by R. Grove, V. Damodaran and S. Sangwan. New Delhi: Oxford University Press.

———. 2006. *Modernising Nature: Forestry and Colonial Eco-Development, 1880–1950*. New York and London: Oxford University Press.

Ramakrishnan, P. S. 1992. *Shifting Agriculture and Sustainable Development: An Interdisciplinary Study from North-Eastern India*. Paris and Park Ridge, NJ: UNESCO and Parthenon.

Rambo, A. T. 1998. 'The Composite Swidden Agro Ecology Stem of the Tay Ethnic Minority of the North-western Mountains of Vietnam'. In *Land Degradation and Agriculture Sustainability: Case Studies from Southeast and East Asia*, Regional Secretariat, edited by A. Patanothai, 43–64. Khon Khen: The Southeast Asian Universities Agro-Ecosystem Network (SUAN), Khon Khen University.

Ramunny, M. 1988. *The World of Nagas*. New Delhi: Northeastern Book Centre.

Rangan, H. 1997. 'Property vs. Control: The State and Forest Management in the Indian Himalayas'. *Development and Change* 28, no. 1: 78–94.

Rangarajan, M. 1996. *Fencing the Forest: Conservation and Ecological Change in India's Central Province, 1860–1914*. New Delhi: Oxford University Press.

Rappaport, R. A. 1968. *Pigs for the Ancestors: Ritual in the Ecology of New Guinea People*. New Haven: Yale University Press.

Rata, E. 1999. 'A Theory of Nontribal Capitalism', *Review* 22, no. 3: 231–90.

———. 2000. *A Political Economy of Nontribal Capitalism*. New York: Rowman and Littlefield.

Reid, R. 1942. *History of the Frontier Areas Bordering on Assam*. Shillong: Assam Government Press.

———. 1944. 'The Excluded Areas of Assam'. *The Geographical Journal* 103, nos. 1 and 2: 18–29.

Report of the Working Group on Forests for the Environment and Forests Sector. 2006. Delhi: GoI Planning Commission.

Report on the Administration of North East India for the Year 1921–22 [1984]. Delhi: Mittal.

Ribot, J. C. 1998. 'Theorising Access: Forest Profits along Senegal's Charcoal Commodity Chain'. *Development and Change* 29, no. 2: 307–41.

Ribot, J. C. and N. P. 2003. 'A Theory of Access'. *Rural Sociology* 64, no. 1: 153–81.

Robben, A. and C. Nordstrom, eds.1995. *Fieldwork under Fire: Contemporary Studies of Violence and Survival*. Berkeley and Los Angeles: University of California Press.

Robben, A. C. G. M. 1996. 'Ethnographic Seduction, Transference and Resistance in Dialogues about Terror and Violence in Argentina.' *Ethos* 24, no. 1: 71–106.

Roe, E. 1991. '"Development Narratives" or Making the Best of Blueprint Development'. *World Development* 19, no. 4: 287–300.

Roque, R. 2010. *Headhunting and Colonialism: Anthropology and the Circulation of Human Skulls in the Portuguese Empire 1870–1930*. New York: Palgrave Macmillan.

Rosaldo, R. 1980. *Ilongot Headhunting 1883–1974: A Study in Society and History*. Stanford: Stanford University Press.

Ross, E. 1998. *The Malthus Factor: Poverty, Politics and Population in Capitalist Development*. London and New York: Zed Books.

———. 2003. 'Malthusianism, Capitalist Agriculture, and the Fate of Peasants in the Making of the Modern World Food System'. *Review of Radical Political Economics* 35, no. 4: 437–61.

Rotenberg, R. 1992. 'The Power of Time and Time to Power'. In *The Politics of Time*, edited by H. J. Rutz. (AAS) American Ethnological Monograph Series No. 4. 18–36.

Rowntree, J. 1981. *A Chota Sahib: Memoirs of a Forest Officer*. Padstow, Cornwell: Tabb House.

Rutz, H. J. and M. B. Erol. 1992. 'Never on Sunday: Time-Discipline and Fijian Nationalism'. In *The Politics of Time*, edited by H. J. Rutz and M. B. Erol. (AAS) American Ethnological Monograph Series, no. 4. New Hampshire: AAS, Washington, DC, 62–85.

Sabarwal, V. K. 1977. 'Science and the Desiccation Discourse on the 20th Century'. *Environment and History* 3: 309–43.

———. 1999. *Pastoral Politics: Shepherds, Bureaucrats and Conservation in the Western Himalaya*. Delhi: Oxford University Press.

Saikia, A. 2011. *Forests and Ecological History of Assam, 1826–2000*. Delhi: Oxford University Press.

Sanga S. 1978. 'Effects of Forest Fallow Cultivation on Forest Production and Soil'. In *Farmers and the Forest: 2 and Marginal Agriculture in Northern Thailand*, edited by P. Kunstadter, E. C. Chapman and S. Sanhasri. Honolulu: University of Hawaii.

Santesombat, Y. 2008. *Flexible Peasant: Reconceptualising the Third World Agricultural Types*. Chiang Mai, Thailand: Faculty of Social Sciences, Chiang Mai University.

Saul, J. and Van Ham, P, eds. 2008. *Expedition Naga: Diaries from the Hills in Northeast India 1921–1937; 2002–2006*. Suffolk: The Antique Collectors Club.

Saur, Carl O. 1972. *Seeds, Spades, Hearths, and Heads: The Domestication of Animals and Foodstuff*. Cambridge, MA: MIT Press.

Schama, S. 1995. *Landscape and Memory*. London: Harper Collins Publishers.

Scheper-Hughes, N. 1992. *Death without Weeping: The Violence of Everyday Life in Brazil*. Berkeley: University of California Press.

Scoons, I. 1999. 'New Ecology and the Social Sciences: What Prospects for a Future Engagement?'. *Annual Review of Anthropology* 28: 479–507.

Scott, J. C. 1976. *The Moral Economy of the Peasant: Rebellion and Subsistence in Southeast Asia*. New Haven and London: Yale University Press.

———. 1990. *Domination and the Arts of Resistance: Hidden Transcripts*. New Haven and London: Yale University Press.

———. 1998. *Seeing Like a State: How Certain Schemes to Improve the Human Condition Have Failed*. New Haven and London: Yale University Press.

———. 2009. *The Art of Not Being Governed: An Anarchist History of Upland Southeast Asia*. New Haven and London: Yale University Press.

Sema, P. 1992. *British Policy and Administration in Nagaland 1881–1947*. Delhi: Scholar Publishing House.

Sharma, J. 2006. 'British Science, Chinese Skill, and Assam Tea: Making Empire's Garden'. *Indian Economic and Social History Review* 43, no. 4: 429–55.

———. 2010. 'Making Garden, Erasing Jungle: The Tea Enterprise in Colonial Assam'. In *The British Empire and the Natural World: Environmental Encounters in South Asia* edited by D. Kumar, V. Damodaran, and R. D'Souza. Delhi: Oxford University Press.

Shneiderman, S. 2010. 'Are the Central Himalayas in Zomia? Some Scholarly and Political Considerations across Time and Space'. *Journal of Global History* 5: 299–310.

Shrivastava, V. 1991. 'The Ethnographer and the People: Reflections on Fieldwork'. *Economic and Political Weekly* 26, no. 24: 1475–84.

Sivaramakrishnan, K. 1999. *Modern Forest: State Making and Environmental Change in Colonial Eastern India*. Stanford and London: Stanford University Press.

Sluka, J. A. 1995. 'Reflections on Managing Danger in Field work: Dangerous Anthropology in Belfast'. In *Fieldwork under Fire: Contemporary Studies of Violence and Survival*, edited by C. Nordstrom and A. C. G. M. Robben, 276–94. Berkeley and Los Angeles: University of California Press.

———. 1990 'Participant Observation in Violent Social Contexts'. *Human Organisation* 49, no. 2, 114–26.

Spencer, J. E. 1966. *Shifting Cultivation in South East Asia*. Berkeley and Los Angeles: University of California Press.

Srinivas, M. N., A. M. Shah and E. A. Ramaswami. 1979. *The Fieldworker and the Field*. New Delhi: Oxford University Press.

Stevenson, H. N. C. 1945. 'The Case for Applied Anthropology in the Reconstruction of Burma'. *Man* 2: 1–4.

Strain, A. and P. Van Ham. 2003. *The Hidden World of the Nagas: Living Tradition in India and Burma*. Munich and New York: Prestel.

Sturgeon, J. G. 2005, *Border Landscape: The Politics of Aka Land Use in Burma and China*. Seattle: University of Washington Press.

Sundar, N. 2011. 'Interning Insurgent Populations: The Buried Histories of Indian Democracy'. *Economic and Political Weekly* 46, 6, 47–57.

Suryanata. K. 1996. 'From Home Gardens to Fruit Gardens: Resource Stabilisation and Rural Transformation in Java'. In *Transforming the Indonesian Upland: Marginality, Power and Production*, edited by T. M. Li, Amsterdam: Harwood Academic Publisher.

Taussig, M. 1987. *Shamanism, Colonialism and the Wild Man: A Study of Terror and Healing*. Chicago: University of Chicago Press.

Thapan, M. 1998. *Anthropological Journeys: Reflections on Fieldwork*. Delhi: Orient Longman.

Thompson, E. P. 1967. 'Time, Work–Discipline, and Industrial Capitalism'. *Past and Present* 38, no 1: 56–97.

Tsing, A. L. 1993. *In the Realm of the Diamond Queen: Marginality in an Out-of-the-Way Place*. Princeton: Princeton University Press.

———. 1996. 'Telling Violence in the Meratus Mountains'. In *Headhunting and the Social Imagination of Southeast Asia*, edited by J. Hoskins, 184–215. Stanford, CA: Stanford University Press.

———. 2002. 'Land as Law: Negotiating the Meaning of Property in Indonesia'. In *Land, Property and the Environment*, edited by J. F. Richards, Oakland, CA: ICS Press.

———. 2008. 'Situating Resource Struggles: Concepts for Empirical Analysis'. In *Contested Grounds, Essays on Nature, Culture, Power*, edited by A. Baviskar. New Delhi: Oxford University Press, 192–216.

Turner, V. 1968. *The Ritual Process: Structure and Anti-Structure.* London: Routledge and Keegan Paul.

Uberoi, P., N. Sundar and S. Deshpandey. 2007. *Anthropology in the East: Founders of Indian Sociology and Anthropology*. Delhi: Permanent Black.

Upadhyay, S. and S. Jain. 2004. *Community Forestry and Policy in North-East India: An Historical and Legal Analysis*. Santa Barbara, CA: Community Forestry International.

Van Schendel, W. 1992. 'The Invention of *Jummas:* State Formation and Nation Building in Southeast Bangladesh'. *Modern Asian Studies* 26, no. 1: 95–128.

———. 2002. 'The Politics of Nudity: Photographs of the "Naked Mru" of Bangladesh'. *Modern Asian Studies* 36, no. 2: 341–74.

Van Schendel, W., M. Wolfgang and K. D. Aditya. 2001. *The Chittagong Hill Tracts: Living in a Borderland.* Dhaka: The University Press Limited.

Van Ufford, P. Q., A. Kumar Giri and D. Mosse. 2003. 'Interventions in Development: Towards a New Moral Understanding of Our Experience and an Agenda for the Future'. In *A Moral Critique of Development: In Search of Global Responsibilities*, edited by P. Q. van Uff ord and A. Kumar Giri. London and New York: Routledge.

Vendergeest, P. and N. Puleso. 1995. 'Territorialisation and State Power in Thailand'. *Theory and Society* 24: 385–426.

Vitebesky, P. 1992. 'Shifting Cultivation Comes to Rest: Changing Values of Land and Person in an Area of Monaragala District'. In *Agrarian Change in Sri Lanka*, edited by J. Brow and J. Weeramunda, 155–87. Delhi: Sage.

Walker, A. and T. Forsyth. 2008. *Forest Guardians and Forest Destroyers: Environmental Politics in Northern Thailand.* Seattle: University of Washington Press.

Whitehead, J. 2010. 'John Locke and the Governance of India's Landscape: The Category of Wasteland in Colonial Revenue and Forest Legislation'. *Economic and Political Weekly* 45, no. 50: 83–93.

Wolf, E. R. 1969. *Peasant Wars of the Twentieth Century*. London: Faber and Faber.

———. 1982. *Europe and the People without History*. Berkeley: University of California Press.

Woodford, R. C. and P. C. Khongwir. 1939. 'How to Save the Soil in the Hills of Assam'. *Assam Agricultural Bulletin*, Vol. 2, pp. 1–3.

Woodthrope, R. G. 1874. Photozincographed at the Surveyor General's Office, Calcutta. Published as black and white plates in 'Rough Notes on the Angámi Nagas and Their Language' by Capt. John Butler in *J. Royal Asiatic Soc. of Bengal*, vol. 44, Part I: 307–46.

Yim, K. 2010. *The Yimchunger and Christianity: A Brief History*. Shamatore: Shamatore Baptist Church.

Zimik, R. 1992. 'Changing Agrarian Structure in Nagaland'. In *Nagaland: A Contemporary Ethnography*, edited by S. M. Chena. New Delhi: Cosmo.

Zimmerer. K. 1994. 'Human Geography and the "New Ecology": The Prospect and Promise of Integration'. *Annals of American Association of Geographers* 841: 150–68.

Zou, D. V. 2005. 'Raiding the Dreaded Past: Representations of Headhunting and Human Sacrifice in Northeast India'. *Contributions to Indian Sociology* 39, no. 1: 75–105.

Zou, D. V. and M. S. Kumar. 2011. 'Mapping a Colonial Borderland: Objectifying the Geo-Body of India's North-east'. *The Journal of Asian Studies* 70, no. 1: 141–70.

Zuliaka, J. 1995. 'The Anthropologist as Terrorist'. In *Fieldwork under Fire: Contemporary Studies of Violence and Survival*, edited by A. C. G. M. Robben and C. Nordstrom, California: University of California Press, 206–23.

INDEX

Note: Page numbers in *italics* and **bold** denote illustrations and tables, respectively. Page numbers with 'n' denote notes.

Lightning Source UK Ltd.
Milton Keynes UK
UKHW04n1810011018
329696UK00001BB/80/P

9 781783 087754